The systematic experiment

The systematic experiment

A guide for engineers and industrial scientists

Edited by
J. C. GIBBINGS
Reader in Mechanical Engineering
University of Liverpool

Contributors:
A. G. BAKER
(Unilever Research Laboratory)

S. L. DIXON, F. DRABBLE, J. C. GIBBINGS, A. K. LEWKOWICZ, D. G. MOFFAT
(Department of Mechanical Engineering, University of Liverpool)

R. SHAW
(Engineering Consultant)

The right of the
University of Cambridge
to print and sell
all manner of books
was granted by
Henry VIII in 1534.
The University has printed
and published continuously
since 1584.

CAMBRIDGE UNIVERSITY PRESS

Cambridge
London   New York   New Rochelle
Melbourne   Sydney

CAMBRIDGE UNIVERSITY PRESS
Cambridge, New York, Melbourne, Madrid, Cape Town, Singapore, São Paulo, Delhi

Cambridge University Press
The Edinburgh Building, Cambridge CB2 8RU, UK

Published in the United States of America by Cambridge University Press, New York

www.cambridge.org
Information on this title: www.cambridge.org/9780521312172

First published 1986
Re-issued in this digitally printed version 2009

A catalogue record for this publication is available from the British Library

Library of Congress Cataloguing in Publication data
Main entry under title:
The Systematic experiment.
1. Science—Experiments. 2. Engineering—
Experiments. I. Gibbings, J. C. II. Dixon,
S. L. (Sydney Lawrence)
Q182.3.S97   1986   001.4'34   85-19061

ISBN 978-0-521-30982-0 hardback
ISBN 978-0-521-31217-2 paperback

Contents

Contents

Preface

Experimenting in science is an exploration, a search into the unknown: this is what makes it so enjoyable. Even more, when it results in the researcher pushing forward the frontiers of knowledge, though to a modest degree, it also gives a sense of achievement.

At seats of learning, instruction in the great majority of branches of engineering and science places a large emphasis upon skill in experiment: and examiners set much import on this ability in the total assessment of the student. And yet, commonly, the student has very limited instruction in the principles of experiment compared with that in what might be called scientific knowledge. We look to the use of this book as a way to correct this marked imbalance.

Though the thrill of discovery may be largely absent from the routine undergraduate experiment, the student can gain, from skilled instruction, an appreciation of the wealth of detail that can be involved in the ordering of an experiment. To assist that understanding this book introduces the totality of the experimental method. It is a distillation of the practical experience of the authors in industry, research establishment and university. It is based on the theme that experiment is logical deduction from observation of a real event and like all logic should be planned and performed in a systematic manner.

This book aims to provide instruction for the undergraduate and to be a reference text for the graduate. Though the applications that it considers make it most suitable for students of engineering, it covers matters of general application for both students and teachers in all branches of science. It is based very largely upon a previous one written mostly by the present

authors and planned and edited by Professor R. K. Penny; his original
initiative and effort is the foundation upon which the present book rests.

J. C. Gibbings
Liverpool
Lent Term 1985

Acknowledgments

We are grateful to the following for permission to reproduce the copyright material listed:

(i) Stress Engineering Services for permission to reproduce the front cover photograph which is referred to in Chapter 5.

(ii) Hodder and Stoughton Ltd and John Farquharson Ltd for permission to reprint the extract from *Gypsy Moth Circles the World* by Sir Francis Chichester, 1967, which is given in Chapter 4.

1

The place of the experiment in engineering

It is the nature of man that he is of an inquiring disposition: the new parent faced with the continual plea of 'Why?' soon learns this. We reveal the features of our material world by observation and we do this most effectively when in an organised way. Experiment embodies organised observation and so it is seen in its position of primacy in man's activity. The principles of organising observation were set forth in detail in the early seventeenth century by Francis Bacon, and his basic ideas have stood the test of time (Ref. [1]). Recently Medawar has written: 'Bacon's writing fired and inspired his readers, and can do the same today. He is still science's greatest spokesman' (Ref. [2]). Sometimes, as in botany, we can make observations and logically deduce classifications without resorting to numerical values: but this is rare. As Kelvin is reported to have said: 'When you can measure what you are speaking about, and express it in numbers, you know something about it, when you cannot express it in numbers, your knowledge is of a meagre and unsatisfactory kind; it may be the beginning of knowledge, but you have scarcely, in your thoughts, advanced to the stage of science' (Ref. [3]). Measurement then is the prime means of experimental observation.

The totality of experiment is more than just ordered observation for it must include logical deduction from what is observed. An experiment must lead to a conclusion even if that conclusion is that the effect sought does not exist; this can still be of value, as in the famous Michelson–Morley experiment that could not detect that our world was moving relative to the light in the space surrounding it.

Our scientific laws are founded on observation: they are formulated to express what we have measured. As Saint-Exupery made one of his characters say: 'Experience will guide us to the rules', he said 'You cannot make rules precede practical experience' (Ref. [4]). Sometimes the evidence is indirect, almost circumstantial. Think, for example, of the intellectual power of Newton in postulating that the gravitational attraction that man had observed to hold us all upon the surface of the earth also acted through the vastness of space (Ref. [5]), so that he thereby explained the orbits of the planets that Kepler had so laboriously computed (Ref. [6]).

Mostly, analysis and experiment go hand-in-hand, as they must do. This is simply because analysis is an ordered description of a real event and determination of the latter is experimentation. Calculation is always preceded by the construction of an analytical model and it is this model that can be tested only by experiment.

Even though scientific laws and analytical models are based on observed values they continually have to be tested by further experiment. For example, a law may be valid only for either the particular conditions or the limited range and scope of the experiment forming its basis. Osborne Reynolds was scornfully critical in this respect when he wrote 'a certain pride in mathematics has prevented those engaged in these investigations from availing themselves of methods which might reflect on the infallibility of reason' (Ref. [7]). When applying laws and analytical models based on experiment to the process of design, the engineer must be particularly careful to ensure that he is fully aware of the limitations of the original experiment and so must have studied it to the necessary extent.

Southwell, a pioneer in the application of calculation by numerical analysis to engineering problems, stressed the importance of understanding experiments from which data is obtained, when he wrote that 'computational methods should allow for the margin of uncertainty which is there in practice whether we like it or not' (Ref. [8]). A calculation is only as accurate as the experimental data fed into it.

This dependence of analysis upon experimental data can be quite critical. It has been found to be particularly so in design studies that seek an optimised solution. An extreme example now stresses this: an attempt to optimise a fleet of oil tankers gave the first computer output as a fleet of three tankers, each 1 mile in length and travelling at 0.5 knots. Refining the data then gave a second output of 1000 tankers, each 2 in long and travelling at 40 knots. The experimental data was being extrapolated too far beyond the existing range of that data. These matters are highly important to the engineer and it is unfortunate that much teaching of analysis is given as an abstract exercise not related to real events and so the young engineer can readily overlook the importance of assessing the realism of an analytical model. It is not a new difficulty; and even the greatest can fail: a translation of the obituary of a great eighteenth-century mathematician reads: 'So scientists have reproached him for having sometimes lavished his calculus on physical hypotheses, or even on metaphysical principles, of which he had not sufficiently examined the likelihood and solidity' (Ref. [9]). The undergraduate course can give the student a false impression that most of engineering practice is straightforward calculation with reliable data.

Only rarely does a teacher describe problems that cannot, or should not, be solved by analysis. Sometimes an analysis is either too complex to be solved by even the largest computers or, if not so, the running cost of the computer is not justified: sometimes, though the equations are known, the boundary conditions are unknown; and sometimes a phenomenon is so complex as not to be understood and so not describable analytically.

There can be occasions when a calculation can be performed but in the time taken for an extensive analysis, followed by preparation of an extended computer program, the equivalent result can be obtained more rapidly and more cheaply by an experiment. This situation can arise particularly in industry during the development of a design. Making a decision at an early stage on which approach to take is an important part of engineering expertise. In engineering design, engineers continually experiment as part of the development process. Sometimes this experiment can be a statistical study of the failures of a product that occur in service.

Experiment is a highly skilled activity and is never to be thought of as the sole province of the scientist: for the engineer, research and development are closely linked and both rest on the fruits of experiment. Because experiment is intellectually demanding it means that, as Parker has said with reference to industrial laboratories: 'A few people of excellence will probably determine which company wins the world-wide race to be first' (Ref. [10]). This is why training in research is so valuable.

In the way that study must be complemented by practice when learning to play a musical instrument, so also is this true for acquiring a skill in experimentation. Perhaps no more is this true than in the ability to report an experiment. For the undergraduate this book can only be of value when its lessons are tested by a carefully arranged set of laboratory exercises: upon the teacher we would urge the greatest care in the choice and arrangement of such exercises. In teaching there are three purposes for experiment. Firstly, there is the experiment used to illustrate a phenomenon described in a lecture. Secondly, there is the laboratory experiment; either this should give exercise in building an analytical model for the event or it should be the basis for a design study. And, thirdly, there is the exercise designed to give experience in the method and techniques of experiment. All three are an important part of a complete course of study in the experimental method.

Returning then to Bacon's views, experiment, like other engineering activities, should be an ordered procedure. It is the purpose of this book to introduce both the design and the organisation of experiment so that it becomes sound engineering practice and consequently economically well justified. The order of the chapters in the book reflect the logical order in which an experiment should be carried out in its entirety. The first step is to plan the experiment in a general way, the second is to consider in more detail how the measurements are to be obtained. The third step is to consider how errors can arise in the measurements and the fourth considers how these errors might be contained. The procedure of experiment must then be thought out. The results have then to be analysed so that both their significance and application can be reasoned. Finally, the transmission of the results must be prepared with care. It is useful to recognise that the procedures set forth here are as valid for the large research programme in a big research establishment as they are for the situation when the superintendent of a group of engineers puts his head round an office door and instructs the occupier to obtain quickly a few extra results from an existing experimental apparatus; or again for the student performing a laboratory experiment: both scales of experiment must be planned for success: planning might take weeks in the former case and five minutes in the latter. The student must develop the habit of pausing for thought before plunging into his set laboratory exercise.

In summary, experiment is an ordered activity which forms a foundation for analytical models of real events. It is these analytical descriptions that the engineer has to create so as to determine his designs: this places experiment in a position of primacy in engineering practice.

References

[1]   Bacon, F. (Lord Verulam) (1868). *The Physical and Metaphysical Works of Lord Bacon* (Trans. J. Devey). Bell and Daldy, London.
[2]   Medawar, P. B. (1979). *Advice to a Young Scientist*. Harper and Row, New York.
[3]   Cole, J. P. (1979). *The Geography of World Affairs*, 5th Edn. Penguin, Harmondsworth, pp. 17–18.
[4]   Antoine de Saint-Exupery (1940). *Night Flight* (Trans. S. Gilbert). Allen Lane, ch. 11.
[5]   Rouse Ball, W. W. (1893). *An Essay on Newton's 'Principia'*. Macmillan, London, pp. 154–6; see also Newton, I., *Philosophiae Naturalis Principia Mathematica*, Prop. 4, Th. 4, 1686 (Trans. Univ. California Press, 1962).
[6]   Koestler, A. (1964). *The Sleep-Walkers*. Hutchinson, London.

[7] Reynolds, O. (1877). 'On vortex motion', *Proc. Royal Inst. Great Britain*, **8**, 272–9, 1875–8 (2 Feb.).

[8] Fox, L. & Southwell, R. V. (1946). 'On the stresses in hooks and their determination by relaxation methods', *Proc. I.Mech.E.* (*App. Mechs.*), **155**, 1–19.

[9] Giacomelli, R. & Pistolesi, E. (1934). 'Historical sketch'. *Aerodynamic Theory* (Ed. W. F. Durand). Springer, Berlin, Vol. 1, Sec. D, p. 325.

[10] Parker, R. C. (1970–71). 'The art and science of selecting and solving research and development problems', *Proc. I.Mech.E.*, **185** (64/71), 879–93.

2
The planning of experiments:
part I – general procedures

2.1 General planning

There are two principle aspects to planning an experiment; one is the
scientific and technical aspect, and the other is the administrative one. To
the engineer, both should be of equal concern, and both are considered here.

2.2 The brief for the experiment

Before any planning is started a brief for the experiment must be drawn up.
Such a brief is rarely, if ever, free of constraints set by others. Usually an
experimenter is under instructions to do the experiment: even if a free agent
in the choice of a subject for investigation there will inevitably be
limitations which, most commonly, are of financial cost.

The onus is always on the experimenter to be quite clear about the brief
that has been set. He must be satisfied both on the adequacy of the
instructions and on the clarity of them. There is an analogy with
architectural practice where the onus is on the architect to ensure a clear
understanding of the client's brief, and this requires not only an initial
briefing, but a continual assessment that both questions and updates the
client's needs.

The brief should contain a statement of the overall object of the
experiment that is comprehensive and unambiguous.

For an undergraduate laboratory class, the object of the experiment is
invariably set quite precisely, because the outcome of the experimental
exercise is known beforehand. For an extended experimental project that

6

an undergraduate might do in a final year of study, the outcome would not be known, at least in full, when the objects of the task were drawn up: it is a research investigation. For a post-graduate research programme, the research student is searching out the unknown, and so the object of the investigation can only be set in general terms: but still it must be set and, in this case, it is important that it be reviewed from time to time as the research proceeds. In industrial practice the experiment can extend from such as a simple test to find the weaknesses in a newly designed component, perhaps, for example, a car door lock, up to a comprehensive fundamental research study of what has been called an 'open-ended' investigation, and which, for example, might lead to the discovery of a new plastic material. For all types of experiment the object must be set before planning begins, even though it is recognised that this may not always be possible with great precision and detail, and also with the recognition that it may be significantly amended as the experiment proceeds, or, indeed, even as the planning continues.

The initial brief must also lay down the principal limitations to the programme. An important limitation is the time available. In industry this is usually laid down precisely. It may be that the results are required by a certain date for a design to be finalised, or it may be that an investigator has only a specific period of time available in which to use a large and intensively worked test plant. This, then, is why it is an important part of the exercise of an undergraduate laboratory experiment that the student complete the investigation in the time allotted. It is important to realise that in engineering practice the time allocated will include the period necessary to complete the investigation by presentation of the results in the required form; the nature of this form is all part of the initial brief: for an engineer in industry on a short investigation, his superior may require only a graph or a table of results; for the post-graduate research student, a complete thesis in a standardised form is required. The preparation of both, however, must be completed within the time set. It can be a salutary exercise for the undergraduate to be required occasionally to produce the results of his experiment in a specified form within the period of his laboratory class.

The other principle limitation to be set in the brief is the financial cost: this includes the cost of the man-power available. For the student this can be represented by the amount of material and equipment available to him; and for the undergraduate by the number of persons in his laboratory group; and for the post-graduate student by the assistance available from technicians. Always some overall limit is set, and often this is separated into a limitation of man-power and a restriction of the other financial costs.

In summary, then, for all types of experiment the researcher must satisfy himself that the brief set is adequately detailed and precisely phrased: these terms of reference must be both suitably comprehensive and unambiguous.

2.3   The information search

The undergraduate will always precede experiment by a search for information: this usually simply consists of reading a prepared document describing his set experiment. The post-graduate research student will be set to read most of the existing literature that is directly relevant to his allotted subject. He will also question his supervisor, and this latter approach in engineering practice in industry is a most valuable one. Even if a professional engineer seeks advice from a colleague in the next office, with suitable past experience, he is gaining information in a most competent and effective way.

A search for information at this early stage, after the brief has been set, is a most important part of research. So much information has been published in the past, that often an item of great relevance has been effectively lost in the multitude of research reports and papers. Time and again there are cases of work being done that turn out to duplicate, sometimes in part, often almost in the whole, work done long ago: such waste of resources and hence of money is bad engineering.

Much can be done to avoid this these days. Enormous data and information banks on computers exist now, even accessible internationally by direct links. Large abstract publications are produced in many disciplines and some that have long existed are now widely available. They include:

(a)   *Engineering Index;*
(b)   *Chemical Abstracts;*
(c)   *Physics Abstracts;*
(d)   *Science Citation Index.*

These four publications are included in the Computer Data Bank of the Lockheed Corporation of the USA. This is on direct access to most large libraries in Europe and North America and it contains some two to three hundred data bases on line.

In addition, for the very latest work, there are lists of contents in recent publications: one such is

(a)   *Current contents*

A most valuable source is the *Science Citation Index*. This lists the authors of works, and those papers that have referred to their works. Thus if a person is known to have written about a topic to be investigated by the experimenter, then the latter can trace other works on this subject. This can often be a valuable way of composing a list of relevant reports.

In preparation for such activity in his later professional career, the student should adopt the laudable practice of doing at least some background reading in his text books before starting a laboratory exercise.

As soon as the graduate enters the engineering profession, he can, usually with great long-term profit, start compiling a subject index to cover the topics of his work. A card index is the most convenient form as it is flexible enough to cope with expansion and changes of subject interest as a career progresses. It must also be flexible in the arrangement of subject divisions to cope with these developments: a simple numbering system is best, but the so-called decimal system, as often used in libraries, whereby each subject level is divided into exactly ten classes, is to be avoided as being quite inflexible; it is a clear example of the tail wagging the dog. The subject index then becomes a useful item in the information search when planning an experiment.

From the briefest questioning to the most extended reading, the information search is an essential prerequisite for assessing how the initial brief might best be met. If, indeed, in industrial practice it results in a decision that the proposed experiment is not needed because the knowledge sought already exists, then this is highly effective engineering practice.

2.4 The scientific plan

The variables to be adjusted in an experiment are to be chosen with care; but it can be equally important, for certainty of the final results, that those variables not controlled are also listed, and the significance of their omission appreciated.

When listing experimental variables, clear distinction must be made between dependent and independent variables. Making this distinction is not always straightforward, and so this matter is discussed in more detail in Section 4.2.

The extent of the list of independent variables controls the extent of the experimental programme. If n experimental readings are to be taken of each of m variables, the total number of readings is $2n^{(m-1)}$: then an experimental programme of this nature can soon get out of control. Even the undergraduate, at this stage in planning the operation of his undergraduate experiment, should assess the amount of data that he is required to collect: on occasions, for example, a student will produce a graph of results in the form of a straight line through an origin with some 15 experimental points on it; this is not competent experimentation.

It is helpful to draw up an order of priority of importance of the independent variables, and this can require considerable physical insight into the phenomenon being studied. Here again, prior knowledge from the information search can be invaluable.

There are two powerful techniques for controlling the extent of an experimental programme for a specified number of variables. These are the use of statistical design of experiments and that of dimensional analysis. Both are described in detail in the two following chapters.

Having listed the independent variables, the ranges of them to be covered by the experiments must be decided. This again has a direct effect on the scale of experimentation through the number of readings to be taken.

Linked with the choice of the number and the range of variables is the distinction between two types of experiment, as follows:

(a)   *The repeatable experiment:* an example of this is a test of the properties of a material within its elastic limit. Or, another example is the test to determine the power output and the heating characteristics of an electric motor.

(b)   *The destruction test:* an example of this is the test to determine the crash-resistant properties of a car. Or, another example is a test to find the arcing damage in an electrical switchgear.

The difficulty with the destruction test is that variables excluded from the list of those chosen for variation might themselves change from test to test. For example, in a simple fatigue test, the successive samples of material might not be identical in composition, and hence in structural behaviour, to the required precision. This uncertainty of composition then becomes an independent variable which is uncontrolled. In this case, statistical design of the experiments can become invaluable.

Often in an experiment a degree of environmental control is required. Certain tests require a temperature control, such as a creep test of materials. Others require a control of the humidity level when, for example, measuring the electrical resistance of plastic materials. Again, it might be necessary to remove dust and other particles from the atmosphere when, for example, using hot-wire techniques for measuring air velocity. It may be found that it is not feasible to gain the required control of the environment: then one may need to include its relevant properties as independent variables.

At this stage in the planning of a large experiment, a preliminary assessment of both the cost and the time schedule of a programme might be drawn up.

In deciding on the variables to be studied, some insight into the physics of the phenomenon to be studied would be used. This, in part, would have come from the prior literature search. Further careful thought at this stage about the physics can be valuable in deciding the correctness of the judgement over the choice and range of variables, but it should certainly be done before undertaking the next stage of planning the experimental equipment: the experimenter must try to judge, as best he can, the likely nature of the physical processes occurring in his experiment.

Thought about the physics of the phenomenon might well show the need for some preliminary, or, as they are called, pilot experiments and their desirability could now be resolved. Such pilot experiments can be invaluable as guides to the final completion of the plan for the full programme. Indeed, they can be found necessary before the initial brief for the full programme is laid down, and the results from these preliminary experiments, often of a comparatively simple nature, can markedly change ideas of the likely knowledge to be gained in a full programme of tests. If the experiments planned are to lead to a design of a product, then a few 'pilot' tests on some individual components of the product can be valuable at an early stage of the experiments.

A study of the physics might indicate the possibility of analogue experiments. One example is in setting up the electrical analogue to complex temperature distributions in a heat-transfer problem. Or again, surface film photoelastic methods can be used for obtaining the distribution of stress in the small part of a complex structure. Such analogue methods of experiment are often found to be invaluable in industrial development work.

It has been mentioned that provision for the number of variables in an experiment can often be markedly reduced through the application of dimensional analysis; but a further application of this analysis, which is also to be detailed in Chapter 4, is to the design of model tests in which the size, or, indeed, other parameters may differ from the full-scale design values. Often only such model tests are feasible, and a decision on the use of this method must be made at this early stage in the planning.

2.5   The design of the experimental rig

The importance of care in drawing up a list of experimental variables has been mentioned. This list has a direct bearing upon the scope, and hence the size and complexity, of the experimental equipment. As well as deciding upon the number of variables, the range of each to be covered in the experiment can greatly affect the rig design.

Again, before a rig can be designed the accuracy required of the experimental results must be judged. This can require considerable knowledge of the physics of the problem, and is greatly helped by previous experience of experiment of a similar kind. Usually it is a decision to be made with care. Firstly, because it can greatly influence the complexity of the rig. Secondly, because design for accuracy of each component must be evenly balanced to give a final comparability amongst the errors of all the variables; it is wasteful to seek great accuracy in one variable if this is to be lost in a much lower accuracy in another component that contributes to the whole.

The lack of reliance on repeatability of the material characteristics in a destruction experiment can sometimes be the overriding factor in the accuracy of an experiment, so that it would be wasteful to design a rig for unnecessary accuracy in other variables such as, for example, the ambient temperature of the experiment. In the case of a chemical reaction, care is always taken to control the purity, and hence the chemical repeatability, of the constituents of the reaction, to the degree originally specified to obtain the required accuracy.

A prior decision to use model tests has an obviously large effect upon the design of a rig, particularly through the resulting determination of the ranges to be covered in the numerical values of the variables.

The design of a rig must be such as to enable a degree of flexibility in its use. This flexibility should be such as to allow both addition of variables and extension of the originally chosen ranges of these variables. Designing for such flexibility is a most difficult matter. It involves decisions which must be based on experienced judgement, and which must, of necessity, often largely be a guess. This is obvious because research into the unknown implies an inability to predict. Unfortunately, such decisions can have marked effects upon the complexity of an experimental rig. The experience that helps such judgement is part of the learning about engineering practice.

In designing a rig, the proposed complexity must always be balanced against the matter of delivery time for both the plant and the materials for its manufacture and use. It can happen that the estimate of an unduly extended time for completion of the rig will require a marked simplification of the design in order to meet the overall time for the experimental programme that was laid down in the original brief. Even if delivery time can be satisfactory, at this stage a preliminary cost estimate for the rig may well show a need for simplification, and indeed a revision of the original brief concerning the list of variables and the ranges of them.

2.6   Choice and design of instrumentation

The primary importance of the care in choice of instrumentation is obvious: extended details of its use are given later in Chapter 5 but, for a full listing of the planning process, some general points are made here.

An initial knowledge of the physics of the problem to be investigated can often show the experimenter whether readings to be taken are going to be steady, fluctuating, or transient in nature. Distinction between the first and the last of these is usually clearly known, but it can sometimes happen that an unexpected fluctuation of reading arises in an experiment, with an important physical significance of the fluctuation. An obvious difficulty then arises if the plan had not provided for the appropriate instrument. This stresses the value of initial careful thought about the physics of the problem to be studied.

It is usually easier to judge both the range and the accuracy of the instruments to be chosen. This choice follows from the careful assessment of the variables, and their range to be covered in the experiment.

The extent to which data is to be automatically recorded is an important decision to be made, if only because of the cost of the provision of such recording. This costing must balance the cost of both the instrumentation and often that of developing the system, both in hardware and computer programming, against the saving in total running costs arising from the more rapid acquisition and processing of data. If development of a system has to be done, then the time taken for this must also be balanced against the later saving of time. There is, however, another important aspect: the experienced researcher is aware that at some point he, or one of his colleagues, will make a human error in recording data. There can even be human error in tabulating or plotting the final results. Being realistic means recognising that human error can only be absolutely eliminated by absolute exclusion of the human element. This cannot be done; it can be greatly reduced by provision of automatic recording, but even then the human element often intrudes through the initial setting of some controls. This matter of human error is returned to again later, but in the initial plan, such may be the extent of the data to be collected that provision of automatic recording becomes a matter of considerable importance in the initial planning.

Associated with the matter of recording data is the topic of interfacing instruments to computers, both to record data and to process it. Here again, human error can intrude in preparation of the computer program. An analogy might be drawn with the advertisement of a food manufacturer that the final product had been untouched by human hand! It might be wondered, in some cases, whether experimental results have not also been unmarked by human intellect. This might seem a light-hearted, or even cynical, remark, but there is a grain of serious concern to be guarded against by the experimenter in planning for automatic recording and assessment where the assessment analysis has been prepared by others: certainly cases have occurred of students using keys on pocket calculators to fit straight lines to experimental points without the least understanding. A real example is shown in Fig. 2.1 for the reader to judge.

In planning the instrumentation, thought must be given to the provision of calibration facilities. A healthy distrust of instruments can usefully be inculcated in the students; a matter discussed again in Chapter 5. The planning for calibration facilities must be given the care merited by the realisation that all the data from instruments, without exception, rests for its accuracy on some form of calibration at some stage. The plan must provide both for initial calibration and for continual recalibration, as needed.

Delivery time predictions for instruments are a part of the planning process, with an importance that is obvious. For the research worker this can also be linked with the programme of allocation of instruments at the conclusion of another experiment preceding the one being planned. This problem of a time schedule for sharing expensive items of instrumentation can arise in an undergraduate extended experimental project; it often occurs for post-graduate research students: it is a common occurrence in industrial experimentation. Often what is a more difficult matter is the judgement of the delivery time of new instruments: experience shows that a generous allowance for late delivery is wise at the planning stage.

An experiment that forms a piece of basic research can often require the development of special instruments when the measurements cannot be

Fig. 2.1. Graph of results showing a least squares fit of a straight line produced by a student using a pocket calculator with a built-in program for this purpose. The experiment was to measure the deflection of a simply supported beam loaded by a weight at the centre of the span. The reader must judge the lack of understanding shown.

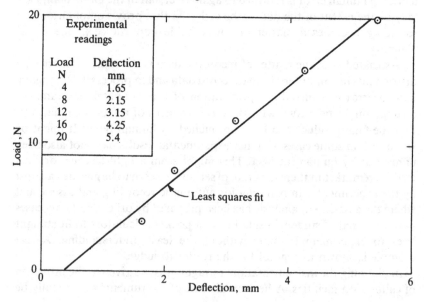

Experimental readings

Load N	Deflection mm
4	1.65
8	2.15
12	3.15
16	4.25
20	5.4

made by ones that are commercially available. This can be a lengthy process, sometimes occupying a significant amount of the time available for the full experimental programme: planning for this then becomes an important matter. Even when an existing measuring instrument is being used it can often happen that a complex rig has to be designed and built just to traverse the instrument over the required positions and to the required precision of location.

The provision of a computer terminal is now a common feature of the provision of instrumentation. When the computer has to be specially interfaced to the measuring instruments, as will be described in detail in Chapter 7, preparation of this can again occupy a lengthy period in the timetable.

2.7 Consumable supplies

The acquisition of consumable materials for an experiment seems rarely to raise problems of purchase, but, rather, of means of supply. For example, electricity supplies might need a special cable, with the associated switch gear, to be laid to the experiment. For certain refined electrical experiments a special earthing system may have to be provided. Supplies of commercial gas for burners or compressed air may have to be piped in, each with suitable valve gear and pressure controllers. The compressed air supplied may need to be of a particular humidity level, which may even have to be variable. Other experiments may require a supply of a suitable inert gas. Some experiments can require large quantities of water, for example, either for flow experiments or for cooling purposes. In the latter case, a choice may have to be made between a direct connection from the public mains supply and a closed circuit system with an attendant heat exchanger for cooling to the atmosphere.

What can often be more troublesome is the provision of suitable storage for consumable materials. For example, a quantity of inflammable liquids, such as fuels, above a certain quantity would require the provision of a fireproof store, perhaps also external to the building housing the experiment. Corrosive liquids, such as acids, again require a storage designed to prevent accidental spillage. Both these cases might require a further provision of containment of an accidental spillage: this can also be required for rigs containing large quantities of water. In an undergraduate experiment, a suitable bench rack for bottles may be required for these safeguards.

Refrigeration may be needed for storage of certain supplies, or for parts of the experimental rig, such as in heat-transfer experiments. On some experiments, storage for supplies of liquid gases may have to be provided. Again, it may be necessary to store certain materials in an atmosphere in which the humidity is carefully controlled.

2.8   Control of safety

It can be difficult to inculcate ideas of safety into undergraduates, simply because this feature has already been taken into account in the preparation of their laboratory experiments: but, in industrial practice, safety of operation is not only of great importance socially, but is a statutory requirement laid upon the engineer.

The matter of safety has already been touched upon in previous sections. It is an integral part of the planning of the experiment, both in the design of the rig, and the proposed conduct of the experiment.

Certain rules for safety are laid down in all industrial nations by central governments, and often also by localised or regional governments. These rules are often relaxed or even do not exist for experiments of fundamental research in comparison with experiments of routine testing, but, where applicable, they can still be stringent. The engineer has to acquaint himself with these rules. Even so, it is important to realise that legal penalties can arise not only when such rules are contravened, with or without unwanted consequences, but, what is probably more important, even when no specific rules are broken but generally accepted and documented standards of practice are not followed.

The anticipation of an accident can be most difficult: common sense combined with imagination and a thorough understanding of the physical processes involved is a great aid. Learning from the long experience of colleagues can be invaluable. For guidance only, a few common causes of mishaps are now listed:

(a) the inadequate briefing of the designers of the rig upon the proposed usage and operating procedures;

(b) the change in the design of a rig by someone who has not been fully briefed on the safety features of the original design;

(c) the modification of the apparatus without an adequate scrutiny of the original design;

(d) the use in the design of safety devices that do not operate as 'fail-safe' devices;

(e) the inadequate briefing of the operators;

(f) an inadequate familiarity with, or preparation of, emergency procedures;

(g) a change in the operating procedure without an adequate briefing of all the operators;

(h) the action possible by unskilled operators not being adequately accounted for;

(i) an undue haste of operation;

(j) a long familiarity with procedure leading to carelessness;

(k) a lack of discipline in operating specified procedures by specified persons;

(l) the removal of a safety device for specific short-term experimental reasons without appropriate changes in operating procedure, and due replacement;

(m) the operation without incident over a long period, engendering a sense of false security;

(n) the assumption that a warning signal arises from a fault in the warning device;

(o) a neglect to repair promptly either a failed safety device, or a faulty warning device;

(p) the inadequate control against access by unauthorised persons.

Achievement of reasonable safety can often require control of the environment. This could involve provision of such things as an inerting system against explosion and a controlled ventilation provision. More protection might be needed by such matters as earthing system, noise defenders and blast protection.

2.9 Administration aspects

Even an undergraduate experiment requires some administration: especially is this so when a group of students have to organise themselves; one may take readings, one may operate controls, whilst another may operate the computer which is programmed to compute the final results, or which in some cases may be interfaced with the experimental instruments.

Administrative arrangements are thus an integral part of planning the experiment, and can, in the engineer's professional practice, include the following topics which are given as examples of the matters that can be involved.

The size of the manpower team becomes an administrative matter, particularly when an experimental programme has to run in parallel with another, so that members of the team have to switch their efforts from one experiment to another. A timetable would then have to be drawn up for this.

Often plant has to be shared with other projects. Examples of this might be high-voltage test plant, gas-turbine or other engine test beds, and large, general-purpose, wind tunnels. This sharing of persons and plant can lead to a research programme of limited flexibility, requiring the test programme to be drawn up carefully in advance.

The provision of services can involve administration of the following matters:

(a) The supply of consumable materials could involve arrangement of suitable rates and intervals of delivery. This might include arranging the supply of test specimens, particularly for destruction testing, from workshops.

(b) The arrangement of computer services for production of results during the testing period covers such matters as organisation of the times of access to the computer, with perhaps the degree of priority of this access to be determined, and also the prior preparation, and, importantly, testing of all the necessary computer programming.

(c) Particularly for large plant, provision of maintenance services has to be planned and timetabled in advance. For the postgraduate student, a tap in a chemical bench rig may require regular cleaning and greasing; a set of electrical slip rings may require regular cleaning and replacement of brushes. Instruments can also require regular maintenance; for example, the probe head of a hot-wire anemometer may require regular cleaning to avoid significant changes in zero reading, and so provision for this should be made in advance of the start of the experiment.

(d) Power supplies and services often have to be arranged. This could include arrangement of electrical, gas and other types of fuel supply, and, in the case of liquid fuels, arrangements for safe storage, already mentioned in Section 2.7, are part of the administrative planning. Other services might be supplies of compressed air, and of water, with attendant arrangements for the latter of suitable sump drainage away from the rig. There can be limitations on the availability of power supplies. For example, even for some large university research equipment, an electrical demand can overload either the university sub-station, or even the supply authorities' local network during periods of peak demand. This can apply particularly for the load imposed during starting-up of a plant: the same can apply to the heavy short-period demand for a water supply, which might cause an unacceptable pressure drop in the supply mains during certain times in the day. A matter not to be overlooked by the engineer planning electrical supplies is that the charge levied by the supply company for big loads is usually strongly controlled by the peak in demand rather than the total electrical energy taken over a period of time.

(e) The calibration of instruments is a matter of great importance already mentioned, and that is considered in detail in Chapter 5. This calibration is often carried out repeatedly during an experiment, and its provision is part of the preliminary administrative planning.

2.10   Critical path planning

The complexities of planning the programme of activities of a large experimental task are greatly eased by use of a so-called 'critical-path' technique. In the following description of this approach, the practices of this method as proposed by Lockyer (Ref. [1]) are largely followed.

The application of critical-path analysis is illustrated here by the specific example of a project which was done in a final year of undergraduate study, but which could well arise in industrial practice. The aim of this project was to rectify the design of a unit consisting of a single-stage, axial flow, fan driven by a variable-speed, electric motor, this unit being situated at the end of a diffuser duct. The fan characteristics were not matched satisfactorily to those of the electric motor, so that the latter was being overloaded whilst the flow distribution at inlet to the fan was not the uniform one for which the blades had been designed.

The first planning step is to list all the separate activities forming the total project. For this project, the listing is now given in Table 2.1. It is important to note that all activities must be entered, including the final examination. Even enforced waiting times must be listed: one such is the time awaiting delivery of the newly designed fan. These items must be listed in a progressive order. Each activity, except the first, before it can be started, may require the prior completion of another activity. Thus, the table must be composed such that for each and every activity those necessary prior activities must precede it in this listing. There is a problem in setting out such a list in that not all activities can be isolated in this way. For example, the information search could be a continual activity throughout the project and, further, its occurrence from time to time in the order of activities might not be predictable.

The next step is to letter the activities in alphabetical order in the list; this is shown in the second column denoted 'activity code'. Following this, the activities are divided into blocks and numbered as in the third column entitled 'block code'. The criterion for choice of a block of activities is that the activities in a single block are to be carried out strictly in succession as listed.

Each activity is now related to the particular activity that is to follow it: if an activity is the final one in a block, then the first activity in succeeding blocks are determined. These following activities are denoted as 'destination codes' in the fourth column of Table 2.1. For example, activity B is to be completed before the start of activity C, whilst activities D, F, I, K and N require the prior completion of activity C.

A first version of the activities network is then drawn: this is shown in Fig. 2.2. Each activity is denoted by an arrow, which is correspondingly lettered, and each stage between activities is shown by a circle. Working from Table

Table 2.1. *Project: re-design of a motor-fan unit*

Job description	Activity code	Block code	Destination code	Stage numbers, start/finish	Man-hours	Duration
Receive brief. Survey existing unit	A	(i)	B	1-2	4	1
Initial information search	B		C	2-3	6	1
Prepare timetable, plan, proposal report	C		D, F, I, K, N	3-4	10	2
Prepare analysis for fan design	D	(ii)	E	4-5	4	1
Prepare computer program for design	E		K	5-9	8	1
Design motor test rig	F	(iii)	G	4-6	6	1
Manufacture and set up test rig	G		H	6-7	6	1
Test motor	H		K	7-9	6	1
Test fan	I	(iv)	J	4-8	10	2
Test duct flow	J		K	8-9	8	2
Design fan	K	(v)	L	9-10	14	3
Build test fan	L		M	10-11	—	8
Test motor and fan unit in duct	M		N	11-12	10	2
Write final report	N	(vi)	O, P	12-13	20	4
Type and prepare report copies	O		Q	13-14	—	4
Prepare oral presentation	P	(vii)	Q	13-14	10	2
Give oral presentation	Q		Q	14-15	2	1

2.1, the network is drawn out in the following stages:

(i)   Activities *A*, *B* and *C* are drawn in succession in accordance with the 'destination codes'.

(ii) Following activity *C*, five arrows are drawn to denote those of *D, F, I, K* and *N*.

(iii) Then activities are continued to be drawn in alphabetical order until *H* is drawn in.

(iv) From the 'destination code' *H* is followed by activity *K*: by adding this, we cover the requirement already drawn that *K* must follow from *C*, and so the original arrow for *K*, developing directly from *C*, is now crossed off, as shown in Fig. 2.2.

(v)   Continuing in alphabetical order, *J* follows from *I*. The 'destination code' for *J* is *K*, and this link is shown dotted in Fig. 2.1. This link is referred to as a 'dummy activity'. Similarly, *E* is also linked to *K*.

(vi) The diagram is continued to activity *N*. Now the previous arrow to denote this activity developing from *C* is superfluous, and so it, too, is crossed off.

(vii) After *O* is linked to *Q* then *P* is also linked to *Q* by a 'dummy activity'.

The preliminary drafting of the network is now completed.

It is now worth re-drawing the network in a neater form. This is shown in Fig. 2.3. In this diagram, the 'dummy activities' have been eliminated in this tidier version. The stages in the network are now numbered as shown in Fig. 2.3 and these numbers are listed in column 5 of Table 2.1.

Fig. 2.2. First version of activities network.

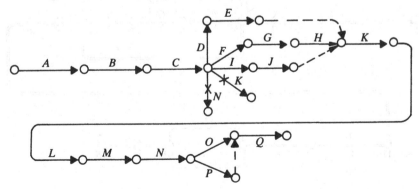

The next stage is to estimate the man-hours and then the duration of time required for each activity. An ability to estimate the man-hours comes from a judgement built upon experience. Even with long experience, such is the nature of research that these estimates can be of doubtful accuracy. A method that is used is to estimate the likely best time, t_a; the likely worst time, t_b; the most likely time, t_m; and to calculate the expected time, t_e, from

$$t_e = (t_a + t_b + 4t_m)/6.$$

This calculation is based on the assumption that the three time estimates form part of a population obeying a β-distribution, and, according to Lockyer (Ref. [1]), can give significantly useful answers. For the present example, the man-hours are given in the sixth column of Table 2.1. Students should take a particular note of the man-hours allowed for writing the final report; commonly they seriously underestimate this item, and frequently this underestimation is the cause of serious delay in submission for a post-graduate degree.

Knowing the man-hours available per week, a likely duration in weeks can be assessed; these are listed in the final column of Table 2.1 and are then entered against each activity in the network of Fig. 2.3.

A prime objective of this planning is to determine the minimum complete time to run the project. In a simple network like Fig. 2.3, this can be seen by inspection, but for a much more complicated network it may not be immediately clear. Thus we now follow a standardised procedure to determine this total duration.

The activities are now arranged in the new listing of Table 2.2 such that the finish stage numbers for the activities are in ascending order. Where

Fig. 2.3. Second version of activities network.

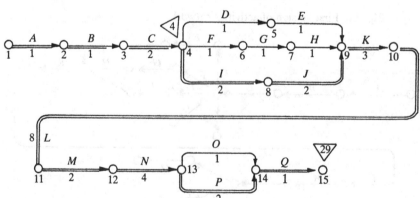

Table 2.2. *Start/finish stage numbers*

Stage numbers			Duration of
Start	Finish	Activity	activity
1	2	A	1
2	3	B	1
3	4	C	1
4	5	D	1
4	6	F	1
6	7	G	1
4	8	I	2
5	9	E	1
7	9	H	1
8	9	J	2
9	10	K	3
10	11	L	8
11	12	M	2
12	13	N	4
13	14	O	1
13	14	P	2
14	15	Q	1

there is more than one activity with a common finish number, then the activities are listed in ascending order of their start numbers. Where there are two activities in parallel, the one of shorter duration is listed first. To this table is added the duration of each activity.

A duration chart is now prepared: this is called a Gantt chart, and is drawn in Fig. 2.4. The procedure is to draw the activities in the order listed in Table 2.2. Each activity is drawn as a line of the corresponding duration, and its start and finish stage numbers are entered on the chart. The start stage number for each activity is placed at the latest finish time for that stage number from a preceding activity. For example, activity 9–10 starts at the conclusion of activity 8–9, and not 5–9 or 7–9.

The minimum total duration of the project is now shown on the Gantt chart as being of 28 weeks. To determine how the activities must be ordered to obtain completion within this minimum time, the so-called 'critical path' is now marked out on the Gantt chart: this is shown by the dotted lines in Fig. 2.4. This path is drawn in Fig. 2.4 commencing from the final stage number 15 and at each step is drawn vertically upwards until it reaches the level of the finish stage number that is the same as that for the start number of the previous activity. For example, the final finish stage 15 has a start stage numbered 14; or the final finish stage 8 has a start stage numbered 4. As well as enabling the determination of the 'critical path', the Gantt chart shows the various degrees of flexibility available in timing some activities.

24   *The systematic experiment*

For example, activity 4–6 could be put back 1 week; activity 5–9 could go back 2 weeks, both without affecting the total critical path time of 28 weeks. Those two periods of respectively 1 and 2 weeks are known as the 'float' available on these activities. The critical path can also be marked out on the network diagram of Fig. 2.3. It is shown by the double line arrows for the corresponding activities.

Certain deadlines can be called for on an experimental programme. In the present example there is one at 4 weeks when the proposal report is required, and another at 29 weeks for the final examination. These are marked on Fig. 2.3 by the triangles with these corresponding duration points entered. The Gantt chart shows that these deadlines would be met.

Fig. 2.4. Gantt chart.

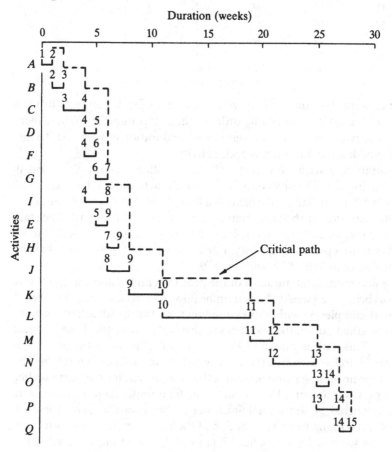

The loading on the staff can now be determined from the Gantt chart, and Table 2.1. The assessment is listed in Table 2.3 and is shown graphically in the histogram of Fig. 2.5. The average man-hours available from the two students working in this project is assessed at 12 per week, and the histogram shows that this is exceeded over three periods. For this task this might be acceptable, for others in industry it might not be so, and then a revision of the plan would be required at this stage. A useful modification can be made by using the float available to move each of activities 5–9 and 7–9 back 1 week. This reduces the high peak load in week 6 on the histogram of Fig. 2.5. The revised loadings are given bracketed in Table 2.3.

This critical-path analysis can readily be extended to costing an experimental programme and details of this are left to further reading (Ref. [1]).

2.11 Reviews of the experimental programme

At this stage a timetable and a total costing of the experiment has been made. Yet at each of the prior stages in the planning of an experiment that have now been described, thought must be given to a reassessment of the previous planning decisions and, indeed, also of the original brief from which all these decisions followed. At the various stages of the plan constraints may be revealed. These could be of cost and of time.

As a result of the complete plan and this final assessment, amendments to the terms of reference may be required. These could, for example, result in

Fig. 2.5. Loading diagram.

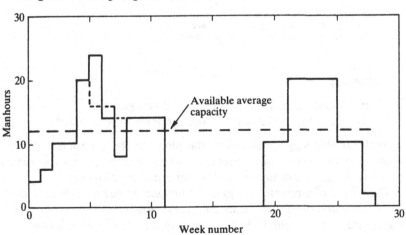

Table 2.3. *Staff loading*

Week number	Man-hours	Total
1	4	4
2	6	6
3	10	10
4	10	10
5	4+ 6+10	20
6	6+10+ 8 (6+10)	24 (16)
7	6+ 8 (8+8)	14 (16)
8	8 (8+6)	8 (14)
9	14	14
10	14	14
11	14	14
12	—	—
13	—	—
14	—	—
15	—	—
16	—	—
17	—	—
18	—	—
19	—	—
20	10	10
21	10	10
22	20	20
23	20	20
24	20	20
25	20	20
26	− +10	10
27	10	10
28	2	2
		Total 260

Notes:
Average man-hours $= \frac{260}{28} = 9.29$ per week.
Man-hours available $= 2 \times 6 = 12$ per week.

the following:

  (i) a change in the variables to be investigated;
  (ii) a change in the range of the variables to be covered.

Again at this stage, and before the plan for the experiment can be finalised, it may be necessary to set up some preliminary 'pilot' experiments to verify the practicability of the original terms of reference.

The nature of a research programme necessarily requires that the plan, and particularly the timetable, should be reviewed from time to time during the course of the research. Even an undergraduate can realise that he is running out of time for the completion of his undergraduate experiment, and so may have to make some quick re-appraisal of what he will be able to complete: this is a good exercise because in engineering practice time schedules have to be kept to rigorously.

2.12   Further general features of planning

In summary, a principal theme of this book has been set in this chapter, which is that careful planning is a vital ingredient for success of the experiment. This applies to any from the simplest task to the large-scale research programme.

Yet all planning must be flexible; that is, there should be both facility, and willingness, to modify as the programme proceeds. It is usual for experimental techniques to improve and even change as the experiment progresses.

Experiment is a creative process, particularly when in engineering practice it is linked to design. Thus in planning an experiment due regard must be paid to human factors, especially when a team is involved, so that researchers may give of their best. This has been considered in detail by several writers, such as Parker (Refs. [2], [3]). Even matters such as ensuring good communication between workers is quite vital for success (Ref. [4]). This can be a problem even in undergraduate laboratory experiments, when students work as members of a group, sharing the duties not only of performing the experiment but also of analysing the results. Yet the engineer is a creative manager, and the management is of people; success in this is the mark of the successful engineer.

Two further special techniques are available to assist the planning. The importance of these justifies the extended discussion of them which follows in the next two chapters.

Finally, perhaps the most difficult decision that an engineer may have to face is the one of stopping an experiment before its planned completion. This can be required, for example, when the contribution to the development of a new design that was hoped for is clearly not being attained. A research worker who is highly motivated in his pursuit of knowledge should not be judged harshly for his opposition to a closure, but rather should receive sympathy for his disappointment arising from his enthusiasm, whilst the decision for stopping is firmly upheld.

References

[1]   Lockyer, K. G. (1969). *An Introduction To Critical Path Analysis*, 3rd Edn. Pitman, London.

[2]   Parker, R. C. (1970—71). 'The art and science of selecting and solving research and development problems', *Proc. I.Mech.E.*, **185**, 879–93.
[3]   Parker, R. C. (1975). 'Human aspects of R and D organization', *Proc. I.Mech.E.*, **189**, 287–94.
[4]   Allan, T. J. (1967). 'Communications in the research and development laboratory', *Tech. Rev.*, **70**(1).

3

The planning of experiments: part 2 – statistical design of experiments

3.1 Introduction

The discussion of the previous chapter has shown that when considering an experimental programme there are five aspects to be considered; these are:

(1)   defining the objective of the programme;
(2)   bringing together existing theory, knowledge and evidence that are relevant to the study;
(3)   setting down the experimental and testing procedures that will be used in order to ensure that the quality of the generated data is as high as is practicable;
(4)   considering the resources available;
(5)   designing the experimental programme.

Although these five aspects appear to be in an ordered sequence, in practice they are so interdependent that they all have to be kept in mind at each stage of the planning. For example, the resources and the range of experimental designs that are available may enable the objective of the programme to be broadened: this chapter is particularly pertinent to this point.

Experimental designs which use 16 experiments (in one case 15) to attain six different objectives are now presented. The designs are simple in concept and their structure is such that in four cases the data generated can be displayed and examined graphically. The reader will find that even if no formal statistical method of analysing the data is used, considerable gains in efficiency will be obtained.

29

The methods for systematically analysing the data which are obtained from the experimental designs are then given. Only arithmetical and graphical methods are used for reaching conclusions. References are given which provide the full statistical procedures for analysing the data. Thus readers whose interest is roused by this chapter can use the references to deepen their understanding of the scope for experimental designs. The references also provide the basis for designs that sometimes involve more and sometimes less experiments.

The basic fact on which the experimental designs are based is that N experiments provide an average and also $(N-1)$ bits of information. The bits of information can be used to measure variability or to measure the effects of changes in processing or formulation variables that are of interest. In the context of experimental programmes, the use of experiments to measure only variability needs to be challenged.

The concept on which experimental designs are based is the generation of a pattern of experiments so that the causative variables which generate a pattern of results can be traced. It is recommended that the reader uses the designs given by first making up results, examining them graphically where possible, and then working through the arithmetical and graphical analysis. In that way, the value of the designs can be assessed with comparatively little effort.

The examples given are not put in a specific context, although some problems in which they would be useful are given. The reason for this is to stress the value of the patterns in the experimental results. Also the designs can be used in many contexts and the experimenter will know the answers to such questions as:

(a) What is the size of an improvement that is of practical significance?
(b) What costs are involved in the various experimental conditions?
(c) How complex an experimental design can be undertaken in a given environment?
(d) Can dimensional analysis be used to reduce the number of experimental variables that have to be studied?

The use of prior knowledge, theory or evidence is discussed. Ideally, forecasts should be made for all the experimental results that will be obtained. Such forecasts can be analysed in the same way as the experimental results. The differences between the experimental and forecasted results can be analysed to assess the gain in knowledge that is obtained from the experimental programme. The analysis of the experimental results can be used to provide better estimates of the experimental results. In that way suspect results can be identified and challenged.

It is assumed throughout that procedures required to produce good experimental data, both from the processing and testing points of view, are used. The importance of these matters cannot be overstressed.

Overall, the acceptance of an experimental result depends on one or more of the following points:

(i) whether it agrees with existing theory;

(ii) whether it agrees with previous experience or evidence;

(iii) whether it fits into an observable pattern;

(iv) whether it is consistent with just random data;

(v) whether the results from repetitions of the experiment are consistent.

The chapter is mainly concerned with the first four points.

3.2 Experimental designs

The basis for the designs is first given and compared with the 'one variable at a time' approach. Experimental designs for six different objectives are then given, along with guidance on the steps required to implement them. Each experimental design uses 16 different experimental conditions.

3.2.1 BASIS FOR EXPERIMENTAL DESIGNS

Comparison with the 'one variable at a time approach'

Fig. 3.1 shows experimental designs for assessing the effects of three processing or formulation variables A, B and C. Each cross on the diagram indicates one numerical value resulting from one test. For example, the experimental programme may be required to find the effects of the three processing variables, pressure, temperature and time on the yield of a chemical reaction. Firstly, a control experiment could be run, then the effect of increasing the pressure can be found, and then the effects of increasing the temperature and time can likewise be found. It is assumed that there are resources available to run eight experiments and that only two levels or types are required to study the effects of changes in A, B and C.

Fig. 3.1(a) illustrates the 'one variable at a time' approach. In this approach, running the eight experiments gives three bits of information about the effects of changing A, B and C. It also gives four bits of information about the processing and testing variability because each test is repeated once.

Fig. 3.1(*b*) illustrates the experimental design based on a cube. This is called a factorial design. In the factorial approach, eight experiments are also run. Seven bits of information are obtained, i.e. the effects of changing *A*, *B* and *C* and the effects of changing *A* and *B*, *B* and *C*, *C* and *A* and also *A* and *B* and *C*. The latter four are called the *AB*, *BC*, *CA* and *ABC* interactions. It is obvious that, whereas the 'one variable at a time' approach only uses four of the results for assessing the effect of changing *A*, the factorial design uses all the eight results for assessing the effect of

Fig. 3.1. Two forms of experimental design: (*a*) one variable at a time; (*b*) *cube based or factorial.* (*The experimental conditions used are denoted by X.*)

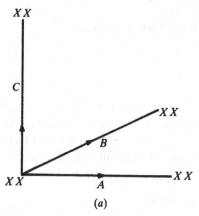

(*a*)

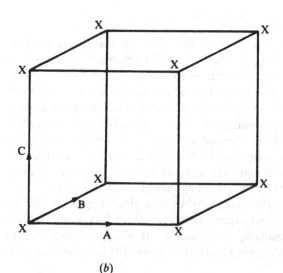

(*b*)

changing *A*, i.e. the four differences between the results at each end of the four parallel edges of the cube, one of which is the *A* axis.

It will be shown later that the factorial approach also gives a measure of the reproducibility of a result.

The advantages of the factorial approach are therefore

(a) It yields more information about the effects of the variables.
(b) It yields more precise information about the effects of the variables.

It will also be shown that the factorial design provides a way of identifying suspect results.

3.2.1.1 First example

The first example now given is to determine the effects of three experimental variables in an experimental region in which there is neither maxima nor minima, in order to test the adequacy of existing theory, knowledge or evidence. The design can also be used to obtain evidence when theory, knowledge or evidence does not exist. The design is given in Fig. 3.2 where illustrative results have been inserted.

The procedure is as follows:

(a) Select two values or types for each of three experimental variables *A*, *B* and *C*. If they are quantitative variables, e.g. temperature and pressure, then choose them to be as far apart as possible. The values will most likely be constrained by practicability and also by the need to avoid maxima or minima.

Fig. 3.2. Experimental design for objective 1, viz. three variables, *A*, *B* and *C*.

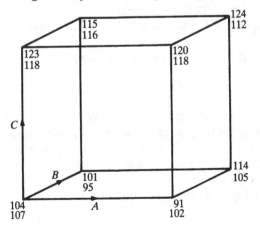

(b) If existing knowledge, theory or evidence is available, provide forecasts of the expected results.

If appropriate, the experiments can be run and the results checked against the forecasts. If the agreement between the results and the forecasts is good, then the programme can be stopped when the evidence of good agreement is conclusive. If appropriate, carry out the two experiments which are considered to give the greatest and least results. This step checks that the experimental region is relevant. The remaining 14 experiments should be carried out in a random sequence (Ref. [1]). The sequence used should be noted.

At this stage, the methodology for analysing the results further could have been given. However, the author considered that the other examples would be more helpful since:

(a) An examination of Fig. 3.2 will give an intuitive appreciation of the effects of each of the variables *A*, *B* and *C*. The intuitive interpretation of experimental results is always recommended and the reader should write down the conclusions reached from such an examination of the results.

(b) The variety of the experimental designs brings out the versatility of the factorial approach.

3.2.1.2   Second example

The second example, now given, is to determine the effects of four experimental variables in an experimental region in which there is neither maxima nor minima, in order to test the adequacy of existing theory, knowledge or evidence. The design can also be used to obtain evidence when theory, knowledge or evidence does not exist.

The design is given in Fig. 3.3 where illustrative results have been inserted.

The procedure is the same as that given for the first example.

3.2.1.3   Third example

The third example, now given, is to determine the effects of five experimental variables in an experimental region in which there is neither maxima nor minima, in order to test the adequacy of existing theory, knowledge or evidence. The design can also be used to obtain evidence when theory, knowledge or evidence does not exist.

The design is given in Fig. 3.4 where illustrative results have been inserted.

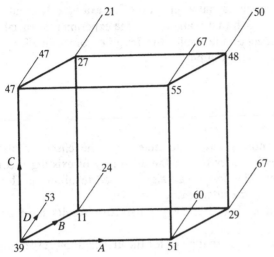

Fig. 3.3. Experimental design for objective 2, viz. four variables, *A*, *B*, *C* and *D*.

Fig. 3.4. Experimental design for objective 3, viz. five variables, *A*, *B*, *C*, *D* and *E*.

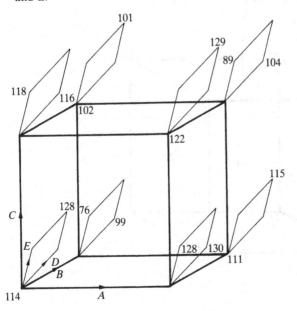

The procedure is the same as that given for the first example. It should be noted that the design is such that if any one of the experimental variables has only a negligible effect on the results, the design becomes, in effect, the same as that for the second example.

3.2.1.4   *Fourth example*

The fourth example, now given, is to determine the effects of three experimental variables in order to test the adequacy of existing theory, knowledge or evidence. The design can also be used to obtain evidence when theory, knowledge or evidence does not exist.

The design is given in Fig. 3.5 where illustrative results have been inserted.

The procedure is the same as that given for the first example, except that more levels of each variable arise.

If the levels of a variable for the cubic or factorial part of the design are L_- and L_+, then the centre level is $(L_- + L_+)/2$. The two axial levels are $(1.1075L_- - L_+)$ and $(1.1075L_+ - L_-)$.

Fig. 3.5. Experimental design for objective 4, viz. three variables, A, B and C with maxima or minima.

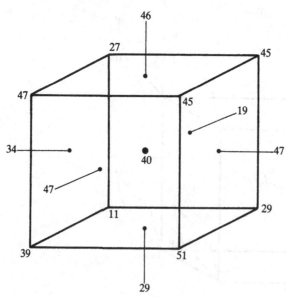

3.2.1.5 Fifth example

The fifth example, now given, is to determine the effects of up to 15 experimental variables in an experimental region in which there is neither maxima nor minima, in order to test the adequacy of existing theory, knowledge or evidence with regard to the key variables. The design can also be used when theory, knowledge or evidence does not exist, to identify the key variables.

An example of a problem in which this problem could be useful is the identification of the key environment variables that affect the stability of a painted surface. The number of the environmental variables is large and a design of this type can help in identifying the key ones so that they can be studied in more detail.

The basis of the design is illustrated in Fig. 3.6 for three variables in four experiments. The diagram has then to be considered in 15 dimensions.

The full design, using −, + to denote the two levels or types of each of the 15 variables, is in Table 3.1, along with the results.

The procedure is the same as that given for the first example. The following should be noted:

(a) The −, + code can be arranged to ensure that at least one of the combinations of variables can be chosen to suit the experimenter.

Fig. 3.6. Basis of experimental designs for objectives 5 and 6, viz. 15 variables – design is shown for three variables; five variables – design is shown for three variables. For objective 5, the number of variables is increased using two levels or alternatives for each. For objective 6, the number of variables is increased and the number of levels or alternatives is also increased.

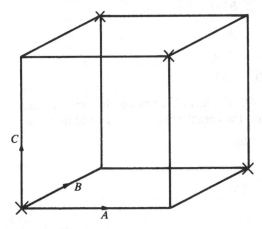

Table 3.1. *Experimental design and results for the fifth example*

A,	B,	C,	D,	E,	F,	G,	H,	I,	J,	K,	L,	M,	N,	O	Results
−	−	−	−	−	−	−	−	−	−	−	+	+	+	−	1
+	−	−	−	+	+	−	+	+	−	−	−	+	−	+	4
−	+	−	−	+	+	+	−	−	+	−	−	−	+	+	12
+	+	−	−	−	−	+	+	+	+	−	+	−	−	−	20
−	−	+	−	+	−	+	+	−	−	+	−	−	−	+	20
+	−	+	−	−	+	+	−	+	−	+	+	−	+	−	25
−	+	+	−	−	+	−	+	−	+	+	+	−	−	−	27
+	+	+	−	+	−	−	−	+	+	+	−	+	+	+	1
−	−	−	+	+	−	−	−	+	+	+	+	−	−	+	38
+	−	−	+	−	+	−	+	−	+	+	−	−	+	−	38
−	+	−	+	−	+	+	−	+	−	+	−	+	−	−	37
+	+	−	+	+	−	+	+	−	−	+	+	+	+	+	30
−	−	+	+	−	−	+	+	+	+	−	−	+	+	−	15
+	−	+	+	+	+	+	−	−	+	−	+	+	−	+	27
−	+	+	+	+	−	+	+	−	−	+	+	−	+	+	50
+	+	+	+	−	−	−	−	−	−	−	−	−	−	−	48

(b)   The first four columns represent the design for the second example.

(c)   The first four columns are in what is called the *standard order*.

(d)   If there are not 15 experimental variables, allocate them in the sequence *ABCD, NMLK*, ... , *E*.

(e)   Each of the following has the complete 16 combinations.

> *A* or *B*, *CDF*
>
> *B* or *C*, *ADG*
>
> *A* or *C*, *BDH*
>
> *A* or *D*, *BCI*
>
> *B* or *D*, *ACJ*
>
> *C* or *D*, *ABK*

Any four from *ABCDE* has the complete 16 combinations.
Each of the following has the complete set of eight combinations:

A or *B* or *C*, *DL*

B or *C* or *D*, *AM*

A or *C* or *D*, *BN*

A or *B* or *D*, *CO*

Table 3.2. *Experimental design and illustrative results for the sixth example*

			A		
		1	2	3	4
B	1	$C_3D_4E_2$, 17	$C_2D_1E_3$, 2	$C_1D_2E_4$, 6	$C_4D_3E_1$, 8
	2	$C_2D_3E_4$, 7	$C_3D_2E_1$, 19	$C_4D_1E_2$, 18	$C_1D_4E_3$, 12
	2	$C_1D_1E_1$, 8	$C_4D_4E_4$, 3	$C_3D_3E_3$, 12	$C_2D_2E_2$, 15
	4	$C_4D_2E_3$, 8	$C_1D_3E_2$, 6	$C_2D_4E_1$, 2	$C_3D_1E_4$, 7

Thus if there are only three variables that are important there is a good chance that the complete eight combinations will be obtained. It is best to use A, B, C and D to cover what are thought to be the more important variables.

(f) The outcome of such an experiment is suggestive rather than conclusive. The conclusions need to be checked by further work. A set or sets of eight or 16 experiments on the lines of the first to fourth examples may be appropriate.

3.2.1.6 Sixth example

The sixth example, now given, is to determine the effects of five experimental variables, in order to test the adequacy of existing theory, knowledge or evidence. The design can also be used to obtain evidence when theory, knowledge or evidence does not exist.

In such a study, the variables are more likely to be qualitative rather than quantitative. Up to four alternatives can be used for each variable.

The basis of the design is illustrated in Fig. 3.6 for three variables in four experiments. The diagram has then to be considered in five dimensions, with four alternatives in each dimension. The design for five variables, each with four alternatives, is given in Table 3.2 and illustrative results have been given.

The experiments are thus, reading in rows,

$$A_1B_1C_3D_4E_2, \quad A_2B_1C_2D_1E_3, \quad A_3B_1C_1D_2E_4,$$

$$A_4B_1C_4D_3E_1, \quad A_1B_2C_2D_3E_4, \quad \text{etc.}$$

Such a design is a fully saturated Hypo–Graeco–Latin Square. The sets of orthogonal squares in (Ref. [1]) provide the designs. The above design was obtained from the 4×4 orthogonal squares, the rows were first randomised and then the columns were randomised.

The procedure is as follows.

Choose the four alternatives for each variable as different as is possible within the relevant area. If there are only three alternatives for one variable, repeat one of them. The steps are then as given in the first example, except of course for the change from two to four variants or levels.

It is to be noted that, by the choice of coding used, at least one of the experimental conditions can be chosen to be a particular combination of variables.

If any three of the experimental variables have negligible effects, the design is such that there will be a complete 4×4 matrix of the other two experimental variables.

3.3   Examination of the experimental results

The examination of the experimental results can provide:

(a) conclusions;

(b) estimates of the experimental results;

(c) checks on the adequacy of previous theory, knowledge and evidence;

(d) guidance as to whether any results are suspect.

Whenever possible, the experimental results should be displayed graphically. The graphical display provides an overall view of the pattern of the results along with an indication of whether the results are consistent.

All the aspects of the examination of the experimental results use arithmetical and graphical methods. Graphical methods are used to show whether there is a systematic pattern in the results or whether the results appear to portray just random variability. Methods are given for estimating the standard deviation of an experimental result.

In an experimental design that a reader uses, the reader would have to supply:

(a) a view of what change in the value of results would be of practical importance;

(b) a view of the costs and benefits that would be associated with any particular experimental condition;

(c) a view on the practicality of using any particular conclusion.

No formal statistical methods of analysis are used since the basis of them would require more space; Refs. [2] and [3] give the formal statistical methodology.

Table 3.3. *Yates analysis of the 2^4 factorial design for the first example*

Conditions				I	II	III	IV	V	VI	VII
X	A	B	C	Results					Effects	Code
−	−	−	−	104	211	396	781	1706	106.6	Average
+	−	−	−	107	185	385	925	−28	−1.7	X
−	+	−	−	83	166	458	11	50	3.1	A
+	+	−	−	102	219	467	−39	20	1.2	XA
−	−	+	−	84	220	22	27	−2	−0.1	B
+	−	+	−	82	238	−11	23	−16	−1.0	XB
−	+	+	−	114	231	−28	9	66	4.1	AB
+	+	+	−	105	236	−11	11	60	3.7	XAB
−	−	−	+	123	3	−26	−11	144	9.0	C
+	−	−	+	97	19	53	9	−50	−3.1	XC
−	+	−	+	120	−2	18	−33	−4	−0.2	AC
+	+	−	+	118	−9	5	17	2	0.1	XAC
−	−	+	+	115	−26	16	79	20	1.2	BC
+	−	+	+	116	−2	−7	−13	50	3.1	XBC
−	+	+	+	124	1	24	−23	−92	−5.7	ABC
+	+	+	+	112	−12	−13	−37	−14	−0.9	XABC

3.3.1 ANALYSIS OF THE RESULTS FROM THE FIRST EXAMPLE

In the experiment suitable levels were chosen for the variables now denoted as *A*, *B* and *C*. The experimental conditions that were thought likely to give the greatest and the least results were run. The results which were obtained were consistent with that thinking and showed also that the range of the results covered the range of interest. The order of the remaining experiments was randomised and the order was noted.

The numerical results are given in Fig. 3.2. For each experimental condition, the first experimental result is the one nearer to the appropriate corner of the cube. An examination of the trends in the results suggests that increasing the value of *C* increases the results whereas changing *A* and *B* has little, if any, effect.

3.3.1.1 *Systematic examination of the experimental results*

X is a pseudo-variable introduced to code the two runs of each experimental condition. X_- denotes the first experiment and X_+ denotes the second one. Thus in effect there are four variables *X*, *A*, *B* and *C*. The steps are now listed and also tabulated in Table 3.3:

(1) List the results in the above sequence, which is the standard order. Note the vertical pattern of minuses and pluses.

(2) Bracket the results in pairs.
(3) Add each pair and enter them in sequence into column II, so that this will fill the upper-half of this column.
(4) From each even result, i.e. 2nd, 4th, 6th, etc., subtract the previous result and enter the differences in sequence into the lower-half of the column II.
(5) Carry out 3 and 4 on column II and enter the results into column III.
(6) Carry out 3 and 4 on column III and enter the results into column IV.
(7) Carry out 3 and 4 on column IV and enter the results into column V.

(Note. Steps 3 and 4 have been done four times for a 2^4 design. In general, steps 3 and 4 are done n times on such a design that uses 2^n experiments.)

(8) Divide the numbers in column v by 16 and enter the results in the effects column VI.

(Note. For a 2^n design, the numbers in column $n+1$ are divided by 2^n.)

(9) Column VII gives the code for the effects, thus X measures half the average difference between the second and first experiments for all the eight experimental conditions, A measures half the average difference between the A_+s and A_-s, etc.

Thus for effects column the first number 106.6 gives the overall average of the 16 results. The second number, -1.7, corresponds as follows:

Average of all the X_- results   Average   Average of all the X_+ results

108.3   106.6   104.9
                                                      etc.

Thus the average of the X_+ results minus the average of the X_- results has the value $2 \times (-1.7) = -3.4$.

The third number, 3.1, corresponds as follows:

Average of all the A_- results   Average   Average of all the A_+ results

103.5   106.6   109.7
                                                      etc.

This method of analysis is known as Yates' analysis and is given in Ref. [4]. The method is given more fully in Refs. [2], [5] and [6]. Useful checks are that the sum of the effects column will equal the last result, 112, allowing for rounding off errors, and that the sum of the squares of each number of the effects column will be $\frac{1}{16} \times$ the sum of the squares of each number in the results column, allowing for rounding off errors. In the general case $\frac{1}{16}$ becomes 2^{-n}.

There are eight experiments which have been repeated. Thus there are eight differences that measure the variability of the results. These eight differences are related to the eight values in column VI that have X in their code. Ignoring the sign, the values in numerical order from this column are:

0.1 (XAC)	0.1 (B)
0.9 $(XABC)$	0.2 (AC)
1.0 (XB)	1.2 (BC)
1.2 (XA)	3.1 (A)
1.7 (X)	4.1 (AB)
3.1 (XC)	5.7 (ABC)
3.1 (XBC)	9.0 (C)
3.7 (XAB)	

The effects of A, B and C from column VI are also written in sequence in numerical order alongside, excluding the first effect which is the average. The eight numbers corresponding to the variability form a group ranging from 0.1 to 3.7. In this group, it could be thought that there is a large gap between 1.7 and 3.1 but a consideration of the codes shows that that is not likely. The seven numbers corresponding to the codes of the experimental variables shows a wider range, from 0.1 to 9.0. It seems reasonable to conclude that changing the value of C, which corresponds with 9.0, has a real effect.

The first conclusion, from the Yates analysis, is that changing C has the effect of changing the response as follows:

C_- gives a result of 97.6, i.e. average of $106.6 - 9.0$,

C_+ gives a result of 115.6, i.e. average of $106.6 + 9.0$.

The standard deviation of results obtained from an experimental condition can be calculated by taking four times (generally $\sqrt{2^n}$) the square root of the average of the sums of the squares of the values in volumn VI whose code contains X, i.e.

$$4X\sqrt{\frac{[(-1.7)^2+(1.2)^2+(-1.0)^2+(3.7)^2+(-3.1)^2+(0.1)^2+(3.1)^2+(-0.9)^2]}{8}}$$

$$= 8.8.$$

3.3.1.2   Estimates of the experimental results

Are there any results which are suspect?

The information in the experimental results appears to be covered by the average, the effect of changing C and the standard deviation. The other effects appear to be measuring the variability. It is always worth checking this point in order to see whether any suspect results are occurring. The best way to do this is to use the inverse Yates analysis (Refs. [3], [6]). The analysis is given below and the calculations shown in Table 3.4:

(1)   The sign codes for the experimental conditions are written down in the reverse standard order.
(2)   The effects that are considered to be real and the average are written down in the reverse standard order.
(3)   Yates analysis is carried out to obtain the estimates.
(4)   The corresponding experimental results are written down, in the reverse of the standard order.
(5)   Each experimental result–estimated result is calculated.

The first useful way to look at the differences or residuals is to plot a histogram. The steps are:

(i)   Calculate range of the differences, i.e. $9.4-(-18.6)=28.0$.
(ii)   Divide the range by 10, i.e. 2.8, and round it off to a convenient number, say 3, which can then be used as the interval for the histogram.
(iii) Define intervals from zero and for each interval construct a column that corresponds to the number of results in the interval as in Fig. 3.7.

The histogram shows clearly that there are four large negative residuals. Thus the experimental process, i.e. the processing and/or testing, is not consistent. The experimental conditions which gave large negative residuals are:

X	A	B	C
+	−	−	+
+	−	+	−
−	−	+	−
−	+	−	−

The large negative residuals do not arise in association with the level of any one variable. The residuals should be plotted in the sequences of the

Table 3.4. *Inverse Yates analysis on the chosen effects of the first example (estimates of results and their differences with experimental results)*

X A B C					Estimates	Experimental Results	Differences
+ + + +	0	0	0	9.0	115.6	112	−3.6
− + + +	0	0	9.0	106.6	115.6	124	8.4
+ − + +	0	0	0	9.0	115.6	116	0.4
− − + +	0	9.0	106.6	106.6	115.6	115	−0.6
+ + − +	0	0	0	9.0	115.6	118	2.4
− + − +	0	0	9.0	106.6	115.6	120	4.4
+ − − +	0	0	0	9.0	115.6	79	−18.6
− − − +	9.0	106.6	106.6	106.6	115.6	123	7.4
+ + + −	0	0	0	9.0	97.6	105	7.4
− + + −	0	0	9.0	106.6	97.6	114	6.4
+ − + −	0	0	0	9.0	97.6	82	−15.6
− − + −	0	9.0	106.6	106.6	97.6	84	−13.6
+ + − −	0	0	0	9.0	97.6	102	4.4
− + − −	0	0	9.0	106.6	97.6	83	−14.6
+ − − −	0	0	0	9.0	97.6	107	9.4
− − − −	106.6	106.6	106.6	106.6	97.6	104	6.4

Fig. 3.7. Histogram of residuals. Design for objective 1.

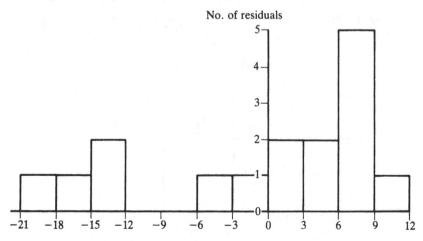

No. of residuals

experiments and testing if they are not the same. The plot(s) will show whether there is any pattern. In this case such a plot showed that there was not any pattern.

The experimental procedures for the inconsistent experimental results have to be checked against those for the consistent ones. If that does not show any correctable mistakes, then the four experiments should be repeated. They were repeated and the results obtained were

(X)	A	B	C	
$(+)$	$-$	$-$	$+$	118
$(+)$	$-$	$+$	$-$	95
$(-)$	$-$	$+$	$-$	101
$(-)$	$+$	$-$	$-$	91

(The X column is now no longer relevant.)

The Yates analysis is repeated and it shows that the original conclusion was valid with only slight changes in values, viz.,

the overall average is 110.3;

average value for C_- is 102.4;

average value for C_+ is 118.2.

The analysis also showed that the variability has decreased. The histogram for the residuals is shown in Fig. 3.8 and it is nearer the symmetrical, centre-based, inverse bell shape that is expected, when the experiments are free

Fig. 3.8. Histogram of residuals. Design for objective 1, with corrections.

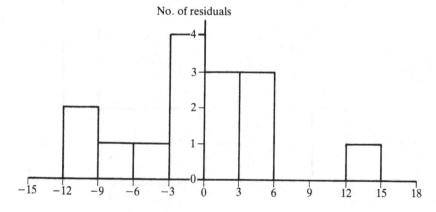

No. of residuals

from any effects other than random variability. The standard deviation has
been reduced to approximately 5.

3.3.1.3 Use of prior theory, knowledge or experience

If there is prior theory, knowledge or experience, and if it can be put in a
quantitative form as follows, then an analysis can be carried out to assess
the adequacy of that prior theory, knowledge or experience.

If it is possible to forecast all the results, then the forecasted results can be
given a Yates analysis and the effects obtained compared with the effects
reached from a similar analysis of the experimental results that are
obtained.

If the size of the effects can be forecast, these can be compared with those
obtained from the Yates analysis of the experimental results. Alternatively
the size of the effects can be put into an inverse Yates analysis and the
pattern of the estimates obtained can be contrasted with the pattern of the
results.

A simple case is used to illustrate some of these points. Suppose that prior
evidence or theory suggests that changing the value of B has little effect. The
effects of A and C are considered to be represented by the following table

	C	
	C_-	C_+
A_-	108	108
A_+	105	121

A

A Yates analysis shows that the effects would be

A	C				Effects	Code
−	−	108 ⎫	213 ⎫	442	110.5	Average
+	−	105 ⎭	229 ⎭	10	2.5	A
−	+	108 ⎫	−3 ⎫	16	4.0	C
+	+	121 ⎭	13 ⎭	16	4.0	AC

The contrast between the prior evidence or theory and the conclusions which have been drawn from the results are thus:

Yates analysis of the experimental results		Conclusions drawn	Prior theory or evidence
	Effects		
A	3.1	0	2.5
C	9.0	9.0	4.0
AC	0.2	0	4.0

The average is around 110 for each case.

The effect of A is compatible, but the effects of C and AC suggest that either

(a) The prior evidence or theory is incorrect.

(b) The experimental methodology has changed, possibly with regard to aspects relating to variable C and/or A. The experimental process would have to be thought through again. It has been noted that there is no time pattern with regard to the residuals.

(c) The materials used in the experimental programme may be different to those from which earlier evidence was obtained.

3.3.2   ANALYSIS OF THE RESULTS FROM THE SECOND EXAMPLE

In the experiment suitable levels were chosen for the variables now denoted as A, B, C and D. The experimental conditions that were thought likely to give the greatest and the least were run. The results which were obtained were consistent with that thinking and showed also that the range of the results covered the range of interest. The order of the remaining experiments was randomised and the order was noted.

Fig. 3.3 shows the experimental results. A visual examination of the results suggests that changing the values of A, B, and possibly D, affects the value of the results.

3.3.2.1   *Systematic examination of the experimental results*

The full Yates analysis is carried out, as in the case for the first example. The analysis is given in Table 3.5. The effects are plotted as a histogram in Fig. 3.9, the single variable and the two variable effects are labelled. The histogram shows the symmetrical, inverse bell-shaped distribution that

Table 3.5. *Analysis of the results of the design for the second example and the estimated results and differences*

A B C D	Results	Effects	Code	Estimates	Differences
− − − −	39	43.5	Average	33.5	5.5
+ − − −	51	9.7	A	52.9	−1.9
− + − −	11	8.9	B	15.7	−4.7
+ + − −	29	3.8	AB	35.1	−6.1
− − + −	47	1.7	C	41.7	5.3
+ − + −	55	0.1	AC	61.1	−6.1
− + + −	27	−0.1	BC	23.9	3.1
+ + + −	48	−1.1	ABC	43.3	4.7
− − − +	53	5.1	D	51.9	1.1
+ − − +	60	2.6	AD	71.3	−11.3
− + − +	24	0.7	BD	34.1	−10.1
+ + − +	67	1.4	ABD	53.5	13.5
− − + +	47	−4.1	CD	43.7	3.3
+ − + +	67	0.2	ACD	63.1	3.9
− + + +	21	−2.7	BCD	25.9	−4.9
+ + + +	50	−1.9	ABCD	45.3	4.7

arises from purely random results, except that the effects for *A* and *B*, and possibly *D* and *CD*, are outside that shape. Confirmatory work may be required for *D* and *CD*. For example, four experiments which permute the two values for *C* and for *D* could be run. In this example *D* and *CD* will be taken as real effects.

The standard deviation of results obtained from an experimental condition can be calculated by taking four times (generally $\sqrt{2^n}$) the square

Fig. 3.9. Histogram of the effects obtained from the experimental design for objective 2.

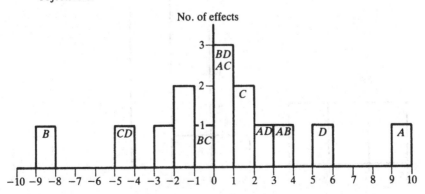

root of the average of the sums of squares of the effects that are not considered to change the results, i.e.

$$4X\sqrt{\frac{[(3.8)^2+(1.7)^2+\cdots+(-2.7)^2+(-1.9)^2]}{11}}=7.6.$$

The terms squared are $AB, C, AC, BC, ABC, AD, BD, ABD, ACD, BCD$ and $ABCD$.

Another way for considering the existence of the effects is to use the half-normal-plot technique given in Ref. [7]. That technique also provides an estimate of the standard deviation. It can be used for analysing all 2^n factorials.

3.3.2.2 Estimates of the experimental results

Are there any results which are suspect?

An inverse Yates analysis is carried out using

A	9.7
B	8.9
D	5.1
CD	−4.1

Fig. 3.10. Histogram of residuals. Design for objective 2.

No. of residuals

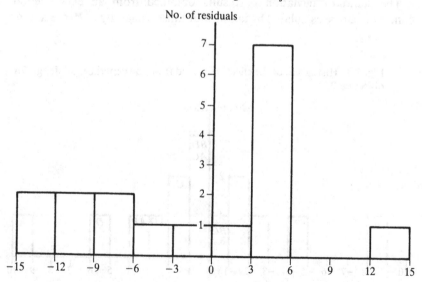

The ensuing estimates of the results has been put alongside the Yates analysis and the differences or residuals calculated. The histogram of the residuals is given in Fig. 3.10. The histogram of the residuals shows that there are no isolated results in terms of numerical value, so no individual result or results can be definitely suspect. However, the shape of the histogram is not consistent with the beginnings of the expected inverse bell shape. In this case a plot of the residuals in the sequence of the experimental order, and testing order if they had differed, showed no pattern. Thus there was no way of identifying any suspect result or results. If the conclusions were critical, a few of the best experimental conditions could be re-run.

3.3.2.3 Use of prior theory, knowledge or evidence

The conclusions can be compared with the prior knowledge as has been given for the first example.

3.3.3 ANALYSIS OF THE RESULTS FROM THE THIRD EXAMPLE

In the experiment suitable levels were chosen for the variables now denoted as A, B, C, D and E. The experimental conditions that were thought likely to give the greatest and the least results were run. The results which were obtained were consistent with that thinking and showed also that the range of the results covered the range of interest. The order of the remaining experiments was randomised and the order was noted.

The numerical results are given in Fig. 3.4. A visual examination of the results is not so easy to assess when all the results, in this case 2^5, are not carried out. In this case only a one-half replicate (16 experiments) of the full 32 permutations of variable levels has been done. A visual examination of the results suggests that increasing the level of A increases the value of the experimental results, whereas increasing the level of B decreases the value of the experimental results.

3.3.3.1 Systematic examination of the experimental results

The experimental results are written down in the standard order for A, B, C and D. The full Yates analysis is carried out, as for the first two examples. The results of the analysis are given below. The codes for the effects are given on the assumption that one and two variable effects are more likely to be real than the three and four variable effects. That assumption is the key to the halving of the full 2^5 factorial. Note the negative signs for $E, AE, BE,$ CE and DE. It is necessary to consult (Ref. [2]) to obtain the basis of these codes. The effects are plotted as a histogram in Fig. 3.11; the effects have

Table 3.6. *Analysis of the results of the third example and the estimated results and differences*

A	B	C	D	E	Results	Effects	Code	Estimates	Differences
−	−	−	−	−	114	111.4	Average	111.3	2.7
+	−	−	−	+	128	4.6	A	127.1	0.9
−	+	−	−	+	76	−11.7	B	80.1	−4.1
+	+	−	−	−	111	0.5	AB	111.5	−0.5
−	−	+	−	+	118	−1.2	C	117.9	0.1
+	−	+	−	−	122	−3.7	AC	120.5	1.5
−	+	+	−	−	102	0.6	BC	102.3	0.3
+	+	+	−	+	89	−3.9	−DE	389.3	−0.3
−	−	−	+	+	128	3.9	D	126.1	1.9
+	−	−	+	−	130	−0.4	AD	127.9	2.1
−	+	−	+	−	99	1.2	BD	94.9	4.1
+	+	−	+	+	115	0	−CE	119.3	2.7
−	−	+	+	−	116	−1.5	CD	118.7	−2.7
+	−	+	+	+	129	3.5	−BE	135.3	−6.3
−	+	+	+	+	101	−0.1	−AE	103.1	−2.1
+	+	+	+	−	104	0.9	−E	104.	−0.1

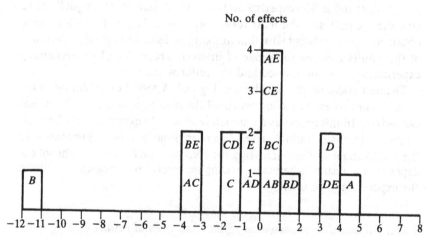

Fig. 3.11. Histogram of the effects obtained from the experimental design for objective 3.

been labelled in order to assist in their interpretation. The histogram shows the beginning of the type of symmetrical, inverse bell-shaped distribution that arises from purely random results except that the effects for *B* and possibly *A*, *D*, *DE*, *BE* and *AC* are outside that shape.

3.3.3.2 Estimates of the experimental results

Are there any results which are suspect?

In order to obtain estimates of the experimental results, and hence be able to see if there are suspect results, all five variables have to be considered. The five variables have to be considered since they all arise in the effects that appear to be real, viz., B, A, DE, BE and AC. An inverse Yates analysis is therefore required for a 2^5 where

$$Average = 111.4$$
$$A = 4.6$$
$$B = -11.7$$
$$D = 3.9$$
$$DE = 3.9$$
$$BE = -3.5$$
$$AC = -3.7.$$

The ensuing estimates of the results have been put alongside the Yates analysis and the differences or residuals calculated. A histogram of the residuals is given in Fig. 3.12. There are no obvious suspect results and so there are not likely to be any results that are seriously misleading. The residuals were plotted in the experimental and testing order if they differed and the plot showed no pattern.

Using this experimental design, the inverse Yates analysis provides estimates of the results from the experiments which are defined by all the 2^5

Fig. 3.12. Histogram of residuals. Design for objective 3.

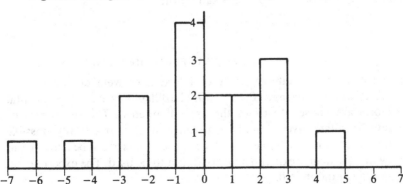

permutations of the values of the variables. Experiments can thus be run to check the estimates of the results for experimental conditions which have not been run. Such experiments provide a good check on the validity of the conclusions.

3.3.3.3   *Use of prior theory, knowledge or evidence*

Again prior theory, knowledge or evidence can be checked, as has already been discussed for earlier examples.

3.3.4   ANALYSIS OF THE RESULTS FROM THE FOURTH EXAMPLE

In the experiment suitable levels were chosen for the variables now denoted as A, B and C. However, in this case there is the central cube, a central point and six axial points. Again the experiments which were thought likely to give the greatest and the least were run. The results which were obtained were consistent with that thinking and showed also that the range of the results covered the range of interest. The order of the remaining experiments was randomised and the order noted.

The numerical results are given in Fig. 3.5. The full method of analysis is given in Ref. [2]. Here the results will only be explored graphically. Fig. 3.13 displays the experimental results. It can be seen that the experimental results obtained by varying A and C give linear relationships, B appears to give a curved relationship. The changes brought about by changing two variables together can be explored, but in this case they did not appear to be important.

If any particular results or experimental regions are of further interest then the appropriate experiments can be run.

3.3.5   ANALYSIS OF THE RESULTS FROM THE FIFTH EXAMPLE

In the experiment suitable levels or alternatives were chosen for the variables now denoted as $A, B, C, \ldots, O$. The differences in the experimental variables are chosen to cover the area of interest. The two levels or qualitative alternatives should be as different as is practically possible within the constraints of relevance. The chosen suffixes can be allocated to allow one particular experimental condition to be used. The experimental sequence is randomised.

3.3.5.1 *Systematic examination of the experimental results*

The experimental results are written down in the standard for A, B, C and D. The full Yates analysis is carried out. The results of the analysis are given in Table 3.7.

It is obvious, even without drawing a histogram, that D and M appear to be the key experimental variables.

Fig. 3.13. Summary of the results from the design for objective 4.

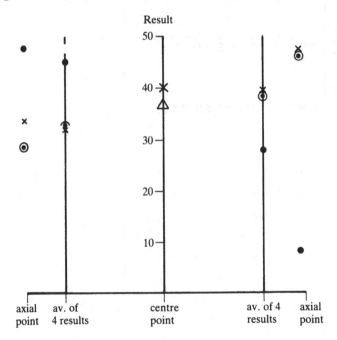

✕ centre point

✗ varying A

● varying B

◉ varying C

△ av. of 8 results from cube

Table 3.7. Analysis of the results of the design for objective 5 and the estimates results and differences

A	B	C	D	E	F	G	H	I	J	K	L	M	N	O	Results	Effects	Code	Estimates	Differences
+	−	−	−	−	−	−	−	−	−	−	+	+	+	−	1	24.6	Average	7	−6
−	+	−	−	+	+	−	+	+	−	−	−	+	−	+	4	−0.4	A	7	−3
−	+	−	−	−	−	+	−	−	+	−	−	+	−	+	12	3.3	B	20	−8
+	−	+	−	−	−	+	+	+	+	−	+	−	+	+	20	−2.9	−F	20	0
−	+	+	−	−	+	+	−	−	−	+	−	−	−	−	20	2.1	C	20	0
+	−	+	−	+	+	−	+	+	+	+	+	−	+	−	25	−0.9	−H	20	5
−	+	+	−	+	−	−	−	−	+	+	+	−	−	−	27	−1.1	−G	7	20
+	−	−	−	−	+	−	−	+	+	+	+	+	+	+	1	1.8	−L	7	−6
−	+	−	+	+	−	−	+	+	+	+	+	−	+	+	38	10.8	D	42	−4
+	−	−	+	−	+	+	−	−	+	+	+	+	−	+	38	0.8	−I	42	−4
−	+	+	+	+	−	+	+	+	−	+	+	+	+	−	37	2.1	−J	29	12
+	−	+	+	−	+	+	+	+	+	−	+	+	+	+	30	0.3	O	29	1
−	+	+	+	+	−	−	−	−	−	−	+	−	−	+	15	−2.4	−K	29	−14
+	−	−	+	+	+	−	+	+	+	−	+	+	−	−	27	3.1	−N	29	−2
−	+	−	+	+	+	−	−	+	−	−	−	−	−	+	50	6.6	−M	42	8
+	−	+	+	−	−	−	−	−	−	−	−	−	−	−	48	−2.7	−E	42	6

3.3.5.2   *Estimates of the experimental results*

Are there any results which are suspect?

The inverse Yates analysis gives:

		Estimates	Experimental code	
			D	M
DM	0	28.8	+	+
M	−6.6	7.2	−	+
D	10.8	42.0	+	−
I	24.6	20.4	−	−

The calculation provides estimates for all the experimental conditions based on the assumption that only the variables D and M matter. The decimal place is rounded off, since it is not relevant. The differences or residuals are reasonable except for the three large differences, viz. 20, 12 and −14. The relevant experiments could be re-run if necessary.

A reasonable experiment to carry out to check that the results and the analysis are meaningful is to set down and run the experiment under those conditions which appear best if all the effects are taken as real, i.e.

$$A_-, B_+, C_+, D_+, E_+, F_+, G_+, H_+, I_-, J_-, K_+, L_-, M_-, N_-, O_+.$$

The last experiment but one in the standard order list gives a result of 50, the greatest result. It differs from the suggested experiment for G, I, K, L and N. Those variables have the corresponding effects 1.1, −0.8, 2.4, −1.8 and −3.1, which are not large. If the experiment given was done, it would show firstly whether the result of 50 was reliable and whether the initial conclusions from the analysis were sufficiently reliable. The experiment was carried out and the result obtained was 53. That result confirmed the value of the experimental approach, the experiments and the conclusions.

If a better result was desired, then one experimental design of the earlier examples would be helpful. The variables could be the quantitative variables which have been shown to be important and others which may not have yet been studied. An EVOP approach (Ref. [3]) may well be the most helpful and cost-effective.

3.3.6   ANALYSIS OF THE RESULTS FROM THE SIXTH EXAMPLE

The experimental variables in this type of study are usually all or mostly qualitative ones, e.g. components in a mixture. The experimental variables were taken as A, B, C, D and E. The experimental sequence is randomised.

The coding of the four alternatives for each variable can be chosen to ensure that at least one of particular interest is carried out, e.g. it may be required to provide a basis for comparison.

The design was:

			A		
		1	2	3	4
	1	$C_3D_4E_2$	$C_2D_1E_3$	$C_1D_2E_4$	$C_4D_3E_1$
B	2	$C_2D_3E_4$	$C_3D_2E_1$	$C_4D_1E_2$	$C_1D_4E_3$
	3	$C_1D_1E_1$	$C_4D_4E_4$	$C_3D_3E_3$	$C_2D_2E_2$
	4	$C_4D_2E_3$	$C_1D_3E_2$	$C_3D_1E_4$	$C_3D_1E_4$

3.3.6.1 Systematic examination of the experimental results

The results obtained and their analysis are given in Table 3.8.

The differences from the overall average can usefully be plotted as a histogram, Fig. 3.14. The 5 × 4 differences are, to some extent, inter-related since each set of four sum to zero. However, provided that is borne in mind, the histogram shows a usable pattern, i.e. there is the usual pattern associated with random variability about the centre and real effects showing outside that pattern.

If the highest result is required then:

(i)   Any A can be used since none of the associated differences break away from the centre pattern. If there are no other relevant criteria, A_4 would be used since it gives the best result.

(ii) B_2 would be used, since it clearly stands out.

(iii) C_3 would be used, since it clearly stands out.

Fig. 3.14. Histogram of differences. Design for objective 6.

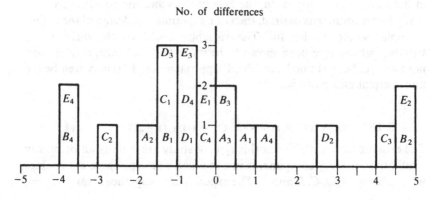

Table 3.8. *Analysis of the results of the design for the sixth example*

		A: 1	A: 2	A: 3	A: 4	Sum	Average	Difference from overall average
B	1	17	2	6	8	33	8.3	−1.1
	2	7	19	18	12	56	14.0	4.6
	3	8	3	12	15	38	9.5	0.1
	4	8	6	2	7	23	5.8	−3.6
	Sum	40	30	38	42	150		
	Average	10	7.5	9.5	10.5		9.4	
	Difference from overall average	0.6	−1.9	0.1	1.1			

The results can also be set out as:

		C: 1	C: 2	C: 3	C: 4	Sum	Average	Difference from overall average
D	1	8	2	7	18	35	8.8	−0.6
	2	6	15	19	8	48	12.0	2.6
	3	6	7	12	8	33	8.3	−1.1
	4	12	2	17	3	34	8.5	−0.9
	Sum	32	26	55	37	150		
	Average	8.0	6.5	13.8	9.3		9.4	
	Difference from overall average	−1.4	−2.9	4.4	−0.1			

The results can also be set out as:

		D: 1	D: 2	D: 3	D: 4	Sum	Average	Difference from overall average
E	1	8	2	8	2	37	9.3	−0.1
	2	18	15	6	17	56	14.0	4.6
	3	2	8	12	12	34	8.5	−0.9
	4	7	6	7	3	23	5.8	−3.6
	Sum	35	48	33	34	150		
	Average	8.7	12	8.3	8.5		9.4	
	Difference from overall average	−0.7	2.6	−1.1	0.9			

   (iv) D_2 would be used, since it stands out. However, D_1, D_3 and D_4 offer useful alternatives.

   (v) E_2 would be used, since it clearly stands out.

If the benefits were additive, $A_4B_2C_3D_2E_2$ would give a result of

$$9.4 + 1.1 + 4.6 + 4.4 + 2.6 + 4.6 = 26.7.$$

Frequently such systems are not additive, but that experiment would be expected to give a result in the middle 20s. The experiment was run and a result of 24 was obtained.

The complete 5^4 of possible experimental results can be estimated in a similar way. When more than one property is measured, experimental conditions which would be expected to give promising results for all the properties can be examined. Ref. [8] gives an example of this type of approach, though not as many variables were used as could have been.

References

[1]   Fisher, R. A. & Yates, F. (1963). *Statistical Tables.* Oliver and Boyd.
[2]   Davies, O. L. (1956). *The Design and Analysis of Industrial Experiments.* Oliver and Boyd.
[3]   Box, G. E. P., Hunter, W. G. & Hunter, J. S. (1978). *Statistics for Experimenters.* John Wiley and Sons.
[4]   Yates, F. (1937). *Design and Analysis of Factorial Experiments.* Imperial Bureau of Soil Science, London.
[5]   Baker, A. G. (1957). 'Analysis and presentation of the results of factorial experiments', *Appl. Stat.,* **6**(1), 45–55.
[6]   Daniel, C. (1976). *Applications of Statistics to Industrial Experimentation.* John Wiley and Sons.
[7]   International Standard. *Statistical Interpretation of Data-Techniques of Estimation and Tests Relating to Means and Variances.* ISO 2854-1976(E). International Organisation for Standardisation.
[8]   Baker, A. G., Carr, P. J. & Nickless, G. (1972). 'Experimental design for exploratory purposes: an example', *Chem. Ind.,* pp. 901–8.

4

The planning of experiments: part 3 – application of dimensional analysis

4.1 The basis and application of dimensional analysis

Dimensional analysis is akin to statistical analysis in that both can often be of use when all attempts to solve the physical laws using what might be called a 'formal' analysis have failed. On its own, dimensional analysis will not provide solutions in the engineering sense of enabling numerical answers to be found. Where it is really powerful is when it is used in conjunction with experiment: this joint use is the theme of this chapter.

Dimensional analysis rests mainly upon a requirement that any functional relation that is an analytical model of a real event is to be so arranged that the terms on both sides of the equation must have an equality of dimension. The logical basis of this requirement is laid down elsewhere (Refs. [1], [2]). This analysis can have the following consequences for experimentation:

(a) It can greatly reduce the amount of experimental investigation by reducing the number of independent variables.

(b) The effect of one variable can be determined by an experimental variation of another.

(c) The applicable range of a variable can be extended beyond the experimental range. In the extreme case this applicable range can be achieved by the measurement of only a single value of each variable.

(d) It can show that sometimes a quantity has no effect upon the phenomena and so can be excluded as an experimental variable.

61

(e) The oversight of an independent variable in the planning of an experiment can be revealed.

(f) The cost of an experiment can be reduced, or even sometimes experimentation can be made feasible, by enabling tests to be made on reduced-scale models of the full-size system. Alternatively it can ease experimental difficulties by enabling experiments to be performed on larger-scale models of very small systems.

This chapter is composed on the assumption that both the method of dimensional analysis and also the pi theorem are familiar to the reader. A very brief summary of this theorem is given in Appendix 1. Discussion here is limited to an explanation and a demonstration of the consequences just described together with others.

4.2 Superfluous variables

For an experiment, care must be taken in distinguishing between dependent and independent variables. An illustration of this is provided by the design of an experiment to determine the power necessary to drive a fan which is pumping air along a duct. This system is sketched in Fig. 4.1. An initial, and perhaps reasonable, expectation could be that this power, P, is a function of the fan diameter, d, the air density, ρ, the air velocity, V, and the fan rotational speed, n. Or, written in standard notation,

$$P = P(d, \rho, V, n). \tag{4.1}$$

But it might be thought that the power depends also upon the rise in pressure across the fan, Δp, so that

$$P = P(d, \rho, V, n, \Delta p). \tag{4.2}$$

This is incorrect. The variable Δp is superfluous for it can be regarded as an alternative dependent variable so that

$$\Delta p = \Delta p(d, \rho, V, n), \tag{4.3}$$

Fig. 4.1. Sketch of an axial flow fan in a duct.

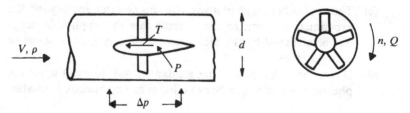

and experiment would confirm this. Thus in equation (4.1) if d, ρ, V and n are specified in value so that from equation (4.3) Δp is fixed, then in equation (4.2) Δp is superfluous as it is not a variable that can be varied in complete independence of all the others.

It can be difficult to assess the physics of a phenomenon sufficiently precisely to determine which of the variables are truly independent. It can often be helpful to visualise what happens in a process. For instance, in this case, once the size of the system, represented by the diameter, d, the air density, ρ, and the duct velocity, V, are set up, then running up the fan to a speed, n, necessarily fixes not only the power, P, but also the pressure rise, Δp, the thrust, T, and the torque Q. However, if the test is on a complete fan and duct system then the duct velocity is no longer an independent variable because running the fan at a speed, n, necessarily fixes the duct velocity, V.

Sometimes the exclusion of such superfluous variables is more obvious. For instance, the fan thrust, T, must not be included as an independent variable because there is the analytical relation

$$P \propto n d T, \tag{4.4}$$

and so again, if d, ρ, V, n are specified, then from equation (4.1) P is fixed and so from equation (4.4) T is determined and is thus not completely independent.

4.3 Avoidance of superfluous experimentation

The application of dimensional analysis, as detailed in Appendix 1, reduces equation (4.1) to a functional relationship between just two variables, each being a non-dimensional group. This relation can be formed as

$$\frac{P}{\rho n^3 d^5} = f\left(\frac{V}{nd}\right). \tag{4.5}$$

If an experiment is designed on the basis of equation (4.1), then a variation of only d would give a single curve on a graph of P plotted against d such as is sketched in Fig. 4.2(a). The additional variation of ρ would result in a family of curves as sketched in Fig. 4.2(b); the variation of V would result in a set of such graphs as shown in Fig. 4.2(c); and finally the variation of n would result in a family of sets of such graphs as seen in Fig. 4.2(d). A descriptive analogy is to liken Fig. 4.2(a), (b), (c) and (d) to, respectively, a page, a chapter, a book and a library. The result is that both the experiment and the results are cumbersome in their extent. But if instead the design of the experiment is based upon equation (4.5) in which, in effect, the five

variables of equation (4.1) have been reduced to two, three quite outstand-
ing benefits ensue:

(a) Only one independent variable, such as, for example, n, needs to
     be varied in the experiment because this results in the required
     variation in the group V/nd.
(b) All the results plot as a single curve on a single graph such as is
     sketched in Fig. 4.3.
(c) The numerical results are independent of the system of units
     used in the experimental measurements because the two groups
     of equation (4.5) are non-dimensional.

An even more striking example is given by the oscillation of a solid in
relation to a point to which it is attached by a spring. If the spring
characteristic is a linear one, so that the extension force is proportional to
the extension through a spring coefficient of proportionality, e, then the
frequency of oscillation, ω, might reasonably be assumed to be a function of
e, the mass of the object, m, and the amplitude of the oscillation, $\mathbf{a}$, or

$$\omega = \omega(e, m, \mathbf{a}). \tag{4.6}$$

Fig. 4.2. A dimensional graph of experimental results for the power required
to drive an axial flow fan.

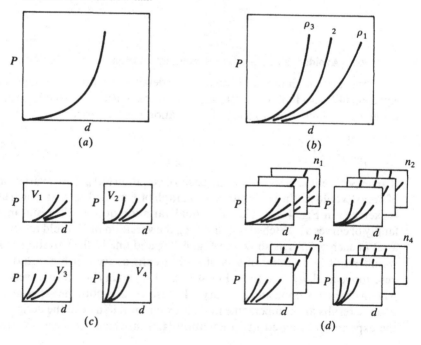

Application of dimensional analysis leads to the particularly compact result of:

$$\frac{\omega^2 m}{e} = \text{constant}. \tag{4.7}$$

This equation shows, without the need of experiment, that the value of **a** has no influence upon the frequency and so it does not need to be included as an experimental variable. Furthermore, dimensional analysis has revealed the nature of the functional dependence of ω upon both m and e. All that is required of an experiment is a test at a single value of each of ω, m and e so that the constant in equation (4.7) might be determined.

The example just given is one for which the analytical solution is easy. But a similar application of dimensional analysis can be made for the vibration of a complex structure. This might be the complete body of a car, maybe the wing and engine structure of an aircraft, perhaps the engine and hull structure of a ship, or the complete structure of a bridge.

The spring coefficient, e, is an elastic force per unit length; Young's modulus, E, is an elastic force per unit area and so if the length scale of the test model is represented by l, the quantity El is the equivalent of e. Again, the vibrating mass, m, can be represented by a density times a volume and so is equivalent to ρl^3 where ρ is the density of the structural material. Equation (4.7) can then be rewritten as:

$$\frac{\omega^2 \rho l^3}{El} = \frac{\omega^2 \rho l^2}{E} = \text{constant}. \tag{4.8}$$

Fig. 4.3. A non-dimensional graph of experimental results for the power required to drive an axial flow fan.

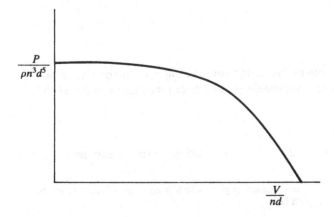

The discussion of the usefulness of equation (4.7) applies to equation (4.8) which is valid for a situation where accurate analysis would not be possible. Equation (4.8) might not always be adequate. For instance, in a complex structure, non-linear spring effects may be present and then the amplitude **a** is of significance and so the group **a**/*l* would be added to equation (4.8) as a further independent variable.

4.4 Number of independent groups

To return to the previous example of flow past a fan, the fan efficiency, η, might also be written

$$\eta = \eta(\rho, n, V, d). \qquad (4.9)$$

If, in addition, the fan shape, as distinct from its size, is varied by altering the angle, θ, of the blades to the plane of the fan,† then

$$\eta = \eta(\rho, n, V, d, \theta),$$

or, using dimensional analysis,

$$\eta = f\left(\frac{V}{nd}, \theta\right). \qquad (4.10)$$

As with a previous example, dimensional analysis has enabled one variable, in this case ρ, to be excluded.

Equation (4.10) might have the form sketched in Fig. 4.4. If only the peak values of the efficiency, η_p, are of interest then a relation between θ and V/nd is implied, as illustrated in this diagram. Thus equation (4.10) reduces to the alternative forms

$$\eta_p = f\left(\frac{V}{nd}\right) \qquad (4.11)$$

and

$$\eta_p = f(\theta),$$

the former being shown as the dotted curve in Fig. 4.4. Further, when only the maximum value of the efficiency is to be determined then equation (4.11) reduces further to

$$\eta_{max} = \text{constant}.$$

This search for certain values such as maxima, minima and zeros is a

---

† It is usual to think of this as a change of pitch which is equal to $\pi d \tan \theta$; hence the designation as a variable pitch fan.

common experimental requirement and, as shown, it reduces the number of non-dimensional groups.

Returning to equation (4.9) this relation reduces to

$$\eta = f\left(\frac{V}{nd}\right) \tag{4.12}$$

and, as before mentioned, only one of V, n and d need to be varied in an experiment. So an experiment over a range of values of V and at a fixed rotational speed, n_1, might give a set of values as sketched in Fig. 4.5. If then the experiment is repeated but at a different rotational speed, n_2, a set of results might be obtained, again as shown sketched in Fig. 4.5, that do not repeat the previous curve. This would indicate that there is an independent variable missing from equation (4.9). In this case it would be the viscosity of air, μ, so that equation (4.12) would be corrected to

$$\eta = f\left(\frac{V}{nd}, \frac{\rho vd}{\mu}\right).$$

The combination of the use of dimensional analysis with an extension experiment in which one extra variable is changed provides this most useful check against the omission of a significant variable.

However, this check might still not reveal a missing variable. For if such a variable had a value of zero or infinity in the experiment then the missing group containing it would have the constant value of zero or infinity. An

Fig. 4.4. A non-dimensional graph of the efficiency of an axial flow fan.

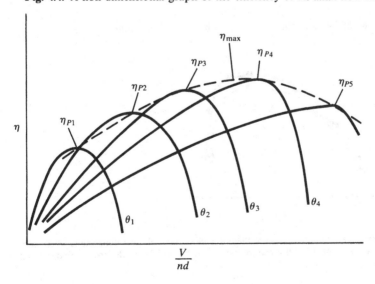

adjustment of another variable would not show the effect of this missing group. This could happen, for example, in an electrical circuit having a very rapid or a very slow initial time decay to an asymptotic state. An example is discussed later in which this time scale could be shorter in the test than in the real situation by a factor of 100:1, it would then be quite possible to miss, in the experimental readings, an initial decay of current which might be important in a real situation.

Sometimes use can be made of a relation basic to the phenomenon being investigated. For instance, if a structure fails under elastic instability, then the failure load, F, can be expressed in terms of Young's modulus, E, and a scaling factor, l, or

$$F = f(E, l). \tag{4.13}$$

If the structure is a single strut and the separate effects of the strut length, l, and the cross-section size, w, are to be determined, then equation (4.13) extends to

$$F = F(E, l, w), \tag{4.14}$$

which becomes†

$$\frac{F}{El^2} = f\left(\frac{w}{l}\right). \tag{4.15}$$

But we know that in bending E appears in the combination EI where I is a representative second moment of area of the strut section. Also

$$I \propto w^4,$$

Fig. 4.5. A non-dimensional graph to reveal a hidden variable.

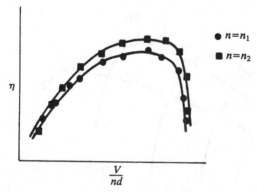

$\bullet\ n = n_1$

$\blacksquare\ n = n_2$

η

$\dfrac{V}{nd}$

† Two groups are obtained because there are effectively only two dimensions which can be taken as L and (M/T^2). This is automatically accounted for by the procedure described in Appendix 1.

so that equation (4.14) becomes

$$F = F(Ew^4, l),$$

giving

$$\frac{F}{El^2} \propto \left(\frac{w}{l}\right),$$

so that

$$\frac{Fl^2}{Ew^4} = \text{constant}, \tag{4.16}$$

where the constant will be a function of the end conditions. Thus the extra information about a fundamental feature of bending results in a reduction of the number of groups from two in equation (4.15) to one in equation (4.16). Experimentation is then required to obtain only the value of the constant in the latter equation.

A different example of the use of extra information is provided by the application of dimensional analysis to the motion of the planets. Suppose the time of orbit, t, is a function of the mass of the planet, m, the gravitational force exerted on the planet by the sun, F, and a size of the orbit, l. Then

$$t = t(F, m, l)$$

or

$$\frac{t^2 F}{ml} = \text{constant}. \tag{4.17}$$

But extra information is available in the form of Newton's law of gravitation which says that

$$F \propto \frac{mm_s}{l^2},$$

where m_s is the mass of the sun and is a constant. Insertion of this relation into equation (4.17) gives

$$\frac{t^2}{l^3} = \text{constant}.$$

This is Kepler's famous law which he derived from extensive observation. Use of the extra information has here enabled a different formulation to be made.†

---

† This example illustrates a way of using dimensional analysis; it does not correspond to the historical development of these relations.

4.5 Effectiveness of experimental variables

Returning to the example of the efficiency of a fan, equation (4.10) shows that if interest is limited to a fan of a single shape so that θ is fixed in value, then

$$\eta = f\left(\frac{V}{nd}\right). \tag{4.18}$$

This reveals another marked usefulness of dimensional analysis because, in an experiment to determine the form of this function, only one of the three parameters V, n and d need be varied. The effect of variation of the other two is then determined without the need for further experimental change of them. Obviously some variables are more conveniently changed than others. In this example a change in d would involve the manufacture of a family of fans, and a change in n would require a variable-speed drive motor which, certainly in large sizes, would be costly in comparison with a fixed-speed one. In contrast, V might more readily be varied over all the finite range illustrated in the sketch of Fig. 4.4 by an adjustable throttling of the flow through the containing duct.

We now consider the important matter of control of variables during an experiment so that it is well designed for correlation of the results. For example, if now viscosity is added to the variables of equation (4.1) then equation (4.5) extends to the two alternative forms:

$$\frac{P}{\rho n^3 d^5} = f\left(\frac{V}{nd}, \frac{\rho V d}{\mu}\right) \tag{4.19}$$

or

$$\frac{P}{\rho n^3 d^5} = f\left(\frac{V}{nd}, \frac{\rho n d^2}{\mu}\right). \tag{4.20}$$

Suppose, firstly, that experimental results are to be represented by equation (4.19). Further, suppose these results are to be shown in the plot of Fig. 4.6(a) which shows a family of curves each of a constant value of $V/(nd)$. This would involve the following difficulty. If the parameters V and n are to be varied and the corresponding values of P measured, then it is not a simple matter to construct these curves. For either by varying V and holding n constant or vice versa the construction of any one of the family of curves is not achieved very readily. But, when the results are to be correlated in the form of Fig. 4.6(b) then a systematic experiment can readily be designed. For holding V constant locates results on a single curve which can then be traced out by varying only n.

Secondly, suppose as an alternative that the experiments are to reproduce equation (4.20). The same difficulty arises in trying to construct the plot of Fig. 4.7(a) for then it is not straightforward to construct any one of the curves using V and n as variables. When the plot of Fig. 4.7(b) is adopted as the basis of the design of the experimental procedure then holding n constant and varying V enables each of the curves to be readily evaluated in turn.

The choice of experimental variable is not always an arbitrary one. This point is now illustrated by two examples.

A phenomenon that has been studied experimentally is that of convection of electrical charge in a dielectric liquid which is set in motion by application of an electric potential to a sharp electrode placed in the liquid (Ref. [3]).

The conduction current, i_c, is found to be a function of the following variables:

    (a)   the electrode spacing, s;
    (b)   the liquid density, ρ;
    (c)   the dielectric coefficient, ε;

Fig. 4.6. A graph to illustrate the control of variables of equation (4.19).

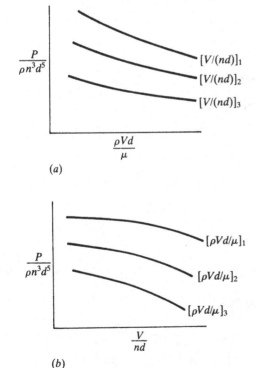

(a)

(b)

    (d)   the liquid viscosity, μ;
    (e)   the liquid conductivity at zero charge density, λ_0;
    (f)   the electrode potential, ϕ.

By dimensional analysis the functional relation can be reduced to:

$$\frac{i_c^2 s^2 \rho^3}{\varepsilon \mu^4} = f\left[\frac{\lambda_0 \rho s^2}{\varepsilon \mu}, \frac{i_c^2 \mu}{\lambda_0 \phi^4 \varepsilon^2}\right]. \tag{4.21}$$

Design of an experiment based upon equation (4.21) might start with the choice of λ_0 and s as being conveniently controlled in the experiment. Inspection of equation (4.21) would suggest that variation of λ_0 and s and measurement of i_c would enable a set of curves to be constructed on a graph, as sketched in Fig. 4.8. This experimental design would be faulty. The reason follows from the proposal to construct the graph whilst holding the variables of ϕ, ε, ρ and μ each at a fixed value. For if we try to construct a non-dimensional group from these four variables we can do so as follows.

Fig. 4.7. A graph to illustrate the control of variables of equation (4.20).

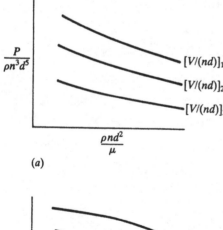

(a)

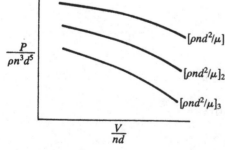

(b)

Such a non-dimensional group will be of the form

$$\phi^a \cdot \varepsilon^b \cdot \rho^c \cdot \mu^d.$$

Setting out the dimensions of these variables gives the dimensions of this group as:

$$\left[\frac{ML^2}{AT^3}\right]^a \left[\frac{A^2T^4}{ML^3}\right]^b \left[\frac{M}{L^3}\right]^c \left[\frac{M}{LT}\right]^d.$$

If this is to be non-dimensional, then for the length dimension it is required that:

$$2a - 3b - 3c - d = 0;$$

for the mass dimension:

$$a - b + c + d = 0;$$

for the current dimension:

$$-a + 2b = 0;$$

and for the time dimension:

$$-3a + 4b - d = 0.$$

The solution of these four equations gives:

$$b = a/2; \quad c = a/2; \quad d = -a.$$

Thus, the non-dimensional group is

$$\left[\frac{\phi^2 \varepsilon \rho}{\mu^2}\right]^{a/2} \quad \text{or} \quad \frac{\phi^2 \varepsilon \rho}{\mu^2}.$$

Fig. 4.8. A graph to illustrate the choice of varied parameters.

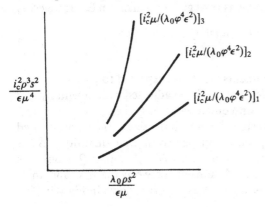

This then is why the experiment would not succeed: a possible non-dimensional group would be held constant so that the full functional relation of equation (4.21) would not be found.

It is now seen that equation (4.21) could be replaced by:

$$\frac{i_c^2 s^2 \rho^3}{\varepsilon \mu^4} = f\left[\frac{\lambda_0 \rho s^2}{\varepsilon \mu}, \frac{\phi^2 \varepsilon \rho}{\mu^2}\right]. \tag{4.22}$$

If the experiment is designed on the basis of either equation (4.21) or equation (4.22) by now varying λ_0 and ϕ whilst holding constant the values of ρ, s, ε and μ, a family of curves could be constructed. For it would not be found possible to construct a non-dimensional group from these latter four variables.

This form of check should always be made. However, it is not always so straightforward and a second example shows when this is so.

Suppose an experiment is planned to determine the rate of heat, $\dot{Q}$, from a vertical plane surface by natural convection to a surrounding gas. This phenomenon has been considered in detail elsewhere (Ref. [4]) where the independent variables are listed as follows:

(a) g, the acceleration due to gravity, and β, the volume coefficient of expansion of the gas, in the combination $g\beta$;
(b) C_v, the coefficient of specific heat of the gas;
(c) k_0, the thermal conductivity, and ρ_0, the density of the gas, in the combination k_0/ρ_0;
(d) μ_0, the viscosity of the gas in the combination μ_0/ρ_0;
(e) l, the size dimension of the plate;
(f) ΔT, the temperature difference between the plate and the gas;

whilst the dependent variable is the combination $\dot{Q}/k_0$.

The discussion already referred to (Ref. [4]) shows that these variables can be reduced to three non-dimensional groups which can be arranged as:

$$\frac{\dot{Q}}{k_0 l \Delta T} = f\left[\frac{g\beta \rho_0^2 C_v^2 l^3 \Delta T}{k_0^2}, \frac{g\beta \rho_0^2 l^2 \dot{Q}}{k_0 \mu_0^2}\right]. \tag{4.23}$$

Following from this, experiments would be expected to plot as a single family of curves. The variables which can be adjusted experimentally and most easily are l and ΔT, the consequent value of $\dot{Q}$ then being measured. So, an experiment could be designed in which the size of the plate is varied and then the heating of the plate is adjusted so that the quantity $l^3 \Delta T$ is held constant. Then the two quantities, $l \Delta T/\dot{Q}$ and $l^2 \dot{Q}$ would be determined, thus obtaining the relationship between the first and third groups in equation (4.23). This might give the curve shown in Fig. 4.9. If an attempt was made to obtain a family of curves by repeating this experiment for other values of $(l^3 \Delta T)$ it would be found that the first single curve was always repeated. The reason can be seen by reformulating equation (4.23) in the form

$$\frac{\dot{Q}}{k_0 l\, \Delta T} = f\left[\frac{g\beta\rho_0^2 l^3\, \Delta T}{\mu_0^2}, \frac{C_v\mu_0}{k_0}\right], \tag{4.24}$$

which shows that the proposed experiment, in which $C_v\mu_0/k_0$ is always held constant, should indeed give only a single curve.

Now there is a difficulty in choosing those parameters to be varied during the experiment. To determine a relationship between the three non-dimensional groups it might be deduced that three of the parameters must be varied whilst four could be held constant. But in this example it is not possible to choose a set of four from the seven and still not be able to form a non-dimensional group from such a set. It would be necessary to vary the type of fluid and this is associated with the fact that in the formulation of equations (4.23) or (4.24) there are four parameters which are properties of the fluid (Ref. [4]): further reading is left to the paper cited.

4.6 Investigation by using scale models

A full-scale test of a system can often be impracticable and then dimensional analysis shows the way by testing on scale models. Suppose it is required to know the steady state voltage distribution in a liquid of low conductivity; the system of interest might be an oil insulated transformer. Then the voltage ϕ could be regarded (Ref. [5]) as a function of a boundary reference voltage, ϕ_0, the coordinate, r, a size of the system, l, the conductivity, λ, the

Fig. 4.9. A graph to illustrate limited independency of fixed parameters.

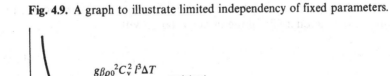

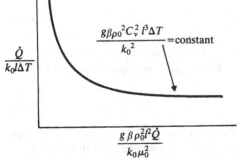

dielectric coefficient, ε, and the coefficient of diffusion of the conducting ions, D. These seven variables form into

$$\frac{\phi}{\phi_0} = f\left(\frac{D\varepsilon}{\lambda l^2}, \frac{r}{l}\right). \tag{4.25}$$

If, in a model test, the independent groups $(D\varepsilon)/(\lambda l)^2$ and r/l each have the same numerical values as the full-scale ones then, whatever the form of the function of equation (4.25), the dependent group ϕ/ϕ_0 must also have the same numerical value at model as at full scale: the form of the function of equation (4.25) does not have to be determined when setting up a test at model scale.

Constancy of the group $(D\varepsilon)/(\lambda l)^2$ means that tests on a reduced size model can only be made if the electrical characteristics of the liquid are changed. It is possible that λ alone could be increased to reduce l in accordance with

$$l^2 \propto \frac{1}{\lambda}. \tag{4.26}$$

For example, a hundredfold increase in λ results in a $\frac{1}{10}$th scale model. Equation (4.25) also shows that a reduced boundary voltage, ϕ_0, can be used in the model test, constance of the group ϕ/ϕ_0 resulting in ϕ values being correspondingly scaled down.

The current density, j, through this type of conductor varies with time, t, after the initial application of the voltage, ϕ_0, so that

$$j = j(\phi_0, l, \lambda, \varepsilon, D, t).$$

This reduces to

$$\frac{jl}{\lambda\phi_0} = f\left(\frac{D\varepsilon}{\lambda l^2}, \frac{\lambda t}{\varepsilon}\right).$$

Now the third group shows that in a model test the time can be scaled by changing the conductivity alone in accordance with

$$t \propto \frac{1}{\lambda}. \tag{4.27}$$

Then the first group shows that the current density can be adjusted by changing one or more of l, λ and ϕ_0 to meet the requirement that

$$j \propto \frac{\lambda\phi_0}{l}. \tag{4.28}$$

To provide a numerical example, for a model conductivity 100 times that of full scale then from equation (4.26) the model would be $\frac{1}{10}$th of full size.

Equation (4.27) shows that a time scale in the model test would be $\frac{1}{100}$th of the full-size scale. The applied electrical field will be proportional to ϕ_0/l. If it is desired to keep this the same in the model test as at full scale then the applied voltage in the model test, ϕ_0 will be $\frac{1}{10}$th of the full-scale voltage. It then follows from equation (4.28) that the model current density will be 100 times that of the full-scale value. Finally, the leakage current, which will be proportional to jl^2, will have the same value in the model test as at full scale. Thus dimensional analysis shows how all the variables are scaled in the model test.

Model tests may be used for a surprising number of phenomena. For example, the problem of thermal effects in solids is often soluble by model tests by analogous arguments to those just given for the electrical phenomena (Ref. [6]). Yet, surprisingly, the many possibilities are little appreciated. The following is an example of a slowly growing awareness (Ref. [7]):

> I learnt the hard way about *Gipsy Moth's* sailing characteristics and so was much interested in Hobbies sailing model of *Gipsy Moth* IV.
>
> When Hobbies made their first prototype hull to scale, it had 'all the appearance of static stability' when the first flotation tests were carried out. When the sails were set and sailing trials began it was found, however, that the keel was not heavy enough, and that the boat tended to heel over on its side, immersing the sails in the water. This is very similar to what occurred when *Gipsy Moth* IV underwent her first sailing trials. The keel was then redesigned and carried aft to the rudder post, just as was prescribed for *Gipsy Moth* in Sydney.
>
> It was also found that the mizzen sail 'tended to tack over and veer the model off course', meaning that the mizzen sail was not balanced by the headsails, and consequently brought the model up into the wind. This same fault in *Gipsy Moth* IV itself was partly overcome in Sydney, when the topmast stays were moved farther forward. This balanced the boat on most points of sailing. In the model, the designers reduced the mizzen sail area, which produced the same effect.
>
> Finally, it was found that the speed and performance of the model were far beyond the makers' expectations. They thought that the hull design and rigging were now the best on the market and when the model's performance was

compared with full radio control racing yachts, it left those behind for both speed and controllability.

My comment on all this is: What a pity that the designers of *Gipsy Moth iv* did not have time to make a model to sail in the Round Pond before the boat was built! What an immense amount of trouble, worry and effort this would have saved me, by discovering *Gipsy Moth's* vicious faults and curing them before the voyage!

References

[1] Gibbings, J. C. (1980). 'On dimensional analysis', *J. Phys.* (*A*), *Math-Gen.*, **13**, 75–89.

[2] Gibbings, J. C. (1982). 'A logic of dimensional analysis', *J. Phys.* (*A*), *Math-Gen.*, **15**(7), 1991–2002.

[3] Gibbings, J. C. & Mackey, A. M. (1981). 'Charge convection in electrically stressed, low-conductivity, liquids, Part 3', *J. Electrostat.*, **11**, 119–34.

[4] Gibbings, J. C. (1981). 'Directional attributes of length in dimensional analysis', *Int. J. Mech. Eng. Educ.*, **9**(3), 263–72.

[5] Gibbings, J. C. (1967). 'Non-dimensional groups describing electrostatic charging in moving fluids', *Electrochimica Acta*, **12**, 106.

[6] Sobey, A. J. (1959). 'Advantages and limitations of models', *J.R.Ae.S.*, **63**, 587.

[7] Chichester, F. (1967). *Gipsy Moth Circles the World*. Hodder and Stoughton, London, Ch. 8, p. 106.

5
Observational and measurement techniques

5.1 Attitudes towards instrumentation

The importance of developing and adopting the correct attitude to instrumentation should be emphasised at the outset. To the engineer, instrumentation is a means to an end, and one should not be overawed by the apparent complexity of equipment. It is not always necessary or advantageous to have a full understanding of the various 'black boxes' employed, although clearly an appreciation of the basic principles is desirable. Equipment should be treated with respect and indiscriminate button pressing should be avoided. Manuals are normally available even on the more basic items, and these should be consulted if necessary.

As a general rule, an initially untrusting attitude towards equipment should be developed, until it is shown from calibration tests and use that this is unwarranted: trust towards another individual comes only with time and experience, and so it is with instruments. Following on from this comes the need for strict honesty in making instrument observations. This should go without saying, but it is a natural tendency to want to ignore faulty equipment or apparently inconsistent results when time is limited. Should an instrument be suspected of malfunctioning, it should be adjusted or replaced, or, if this is not possible, the likely error involved should be acknowledged. Admission of limitations adds credence to reported experimental results.

5.2 Equipment selection

The basic function of engineering instrumentation is to convert physical quantities, which are not normally themselves directly measurable, into quantities which are. Invariably the ultimate quantity employed is linear (or angular) displacement. For example:

U-tube manometer:	change in pressure – change in fluid level.
Thermocouple:	change in temperature – change in emf – needle rotation on voltmeter.
Spring balance:	change in load – deflection of spring.
Pressure transducer:	change in pressure – deflection of diaphragm – change of strain – change in emf – scale change on recording equipment.

There are many factors involved in the choice of equipment, not least of which are *range* and *accuracy*. Clearly, the two are related in that accuracy will be reduced if the portion of the total instrument range used is small. Normally the range to be measured will be known within reasonable bounds, and a sensible choice of instrument can be made. For example, whether to use a U-tube manometer or a Bourdon tube type pressure gauge; whether to apply load by means of dead weights or using a tensile testing machine; whether to measure dimension changes with a micrometer or a metre rule, etc. *Accuracy* should not be confused with *precision*: accuracy is concerned with absolute correctness of measurement, whereas precision entails consistency, i.e. repeatability and stability with time (see Chapter 8, Section 8.1, 'accuracy and precision').

Other factors encountered in the choice of instruments are:

1. *Sensitivity*, defined as the ratio

$$\frac{\text{index movement}}{\text{corresponding change in measured quantity}}.$$

2. *Discrimination* (or resolution), defined as the smallest change in the measured quantity that can be detected. This assumes importance in 'null-balance' methods, examples of which are Wheatstone bridge techniques and micromanometers.

3. *Dynamic response*, which is to do with the ability of an instrument to respond accurately to a rapidly changing input. Systems where this assumes importance are, for example, the U-tube manometer, moving coil galvanometers and various other more complex recording instruments. The response of such instruments can be approximated by the single degree of freedom, second-order, oscillating spring system shown schematically in Fig. 5.1 and represented by the differential equation,

$$m\frac{\mathrm{d}^2x}{\mathrm{d}t^2} + c\frac{\mathrm{d}x}{\mathrm{d}t} + kx = \text{applied force } F(t), \tag{5.1}$$

where m is the mass of the body; c the damping coefficient of the dashpot; and k the spring stiffness. The following analogies can be drawn:

Instrument inertia – inertia force $\left(m \dfrac{d^2 x}{dt^2} \right)$;

Instrument resistance – damping force $\left(c \dfrac{dx}{dt} \right)$;

Instrument stiffness – spring force (kx);

Instrument signal – exciting force $F(t)$;

Instrument deflection – displacement (x).

Fig. 5.2 illustrates the categories of response that can be obtained when the instrument signal is applied instantaneously and then maintained constant. For situations where damping can be controlled it is common to have slight underdamping which tends to improve reading accuracy.

4. *Interference* with the experimental set-up. Examples of this would be the reinforcing effect of a strain gauge on a thin plate test specimen, the effect of a flow measuring device inserted into a pipe-line, or lead resistance effects in electrical circuits.

All of the above factors may have to be considered in selecting the correct instrumentation for the job to be done. Of course, the student will often find that the choice of equipment for his laboratory exercise has already been

Fig. 5.1. Oscillating mass system.

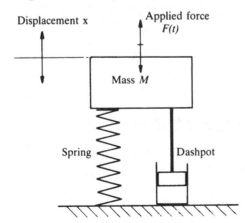

made, or that in the case of projects the most suitable instrument is not available. This is often unavoidable, but an appreciation of the above factors will assist in the assessment of results, and constructive criticism of experimental apparatus is always acceptable in laboratory reports. In any case it is useful, at an early stage, to learn to improvise when the quality of available equipment is limited. This is invariably the case when real problems are tackled with limited time and finances.

5.3   Calibration

Calibration is the comparison between one instrument and another known to be more accurate, under conditions as nearly as possible identical to those existing in the test set-up. The basic standards for the six fundamental quantities, length, mass, time, electric current, temperature and luminous intensity, are controlled, in so far as Great Britain is concerned, at the National Physical Laboratory, Middlesex, and in the United States of America at the National Bureau of Standards. From these basic standards, local standards are produced from which calibrations can be conducted as required. Clearly, the greater the number of calibration 'steps' from the basic standard, the less reliable will be the instrument in question.

Strictly speaking, check calibrations should be conducted before every test series, but where this is not possible preliminary runs should be conducted to give a 'feel' for the equipment involved. Due allowance must be made for possible inaccuracies. Records of instrument calibrations are often filed as a matter of routine and, as a first step, these can be checked and used if they are up to date.

Sources of error in experimentation are dealt with in detail in Chapter 8 and specific instrument errors commonly encountered are included later in

Fig. 5.2. Damping characteristics.

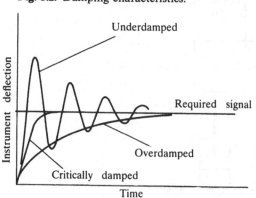

the present chapter. However, in general, apart from the more obvious sources like incorrect zero setting, damage to indicating mechanisms, etc., there are a number of contributory causes to variations of measuring instruments with time, among which are ageing of springs, wear in joints and friction effects. Examples of these can occur in micrometers, dial gauges, pressure gauges and stop-clocks. The net result is often to cause a hysteresis effect in the instrument output which results in different readings depending on whether the required value is approached from below or above.

The importance of identical calibration and operating conditions has already been emphasised. One example illustrating this is the misuse of dial gauges for deflection measurement. These gauges are calibrated to measure movement in the axial direction of the spindle and, clearly, if the dial gauge is not set up with the spindle in line with the required movement, then errors, usually termed 'cosine errors', will be introduced. For example, a 10° misalignment will give an error of 0.15 mm in 10 mm. There is little point in using an instrument capable of reading to within 0.01 mm if the correct operating conditions are not met.

Where more sophisticated electronic equipment is involved, errors are often introduced by not complying with the specified 'heating-up' period. This can in some cases be of the order of, say, 10 minutes during which time apparent results could vary. Another important cause of errors in electrical equipment is that of interference effects from adjacent equipment. Techniques for dealing with this include careful earthing and the use of screened leads, but otherwise the problem is beyond the scope of this text.

5.4 Summary

The foregoing can be summarised in three prerequisites in the use of equipment and instrumentation, if the best results are to be obtained:

(a) the need to develop an initially untrusting but respectful attitude to instrumentation at large;

(b) the importance of correct equipment selection for the job on hand;

(c) the need to ensure through careful calibration that the results obtained are meaningful.

5.5 A compendium of engineering laboratory equipment

In the following sections an attempt is made to outline the various categories of instrumentation that are likely to be encountered in an undergraduate engineering course. The principles of operation are briefly explained and accuracy and range of operation discussed. It is not claimed that the list is complete; the variety of equipment and instrumentation available nowadays precludes a fully comprehensive survey and, where further details are required, reference can be made to one of the many excellent textbooks available on the subject, some of which are listed at the end of this chapter.

5.5.1 MEASUREMENT OF LENGTH

Strictly speaking, the measurement of length is only one aspect of the general field of mensuration which includes measurement of length, angle, area and volume, but since the latter three are derived from length measurements we will be concerned here mainly with length. The field of metrology, i.e. the application of mensuration methods to workshop practice, is not covered in detail, this being a study in its own right, e.g. the excellent manual from the British Standards Institution, Brooker (Ref. [6]). Mention is made below of the more basic workshop tools.

For general dimensioning of components, *wood or steel rules* are normally satisfactory. Wood has a smaller thermal coefficient of expansion than steel, which can be an advantage, but, on the other hand, steel rules are more readily engraved for greater precision and are not so susceptible to end wear. Nominal dimensions, e.g. tube and shaft diameters, are normally measured by means of machinists' *calipers* used in conjunction with a steel rule. Surprising accuracy can be achieved with this relatively crude method. For example, it is quite possible to measure the diameter of a shaft, consistently to within ± 0.2 mm of its true diameter, in this way.

For more accurate dimensioning, use is made of a *vernier caliper* or a *vernier height gauge*. By means of the vernier scale an additional significant figure in the measurement can be obtained over conventional rules.

Suppose a dimension is required to the nearest 0.1 mm using vernier calipers. The vernier and main scales would be as shown in Fig. 5.3. The ten divisions of the vernier scale are spaced over 0.9 cm of the main scale and the aim is to select the vernier graduation which is nearest to being in line with a main-scale graduation. In the sample setting in the illustration, this appears to be the digit 3 and the required dimension is thus 20.3 mm.

The micrometer caliper or simple *micrometer* is, as the name suggests, a direct-reading caliper arrangement, the constituent parts of which are shown in Fig. 5.4. Accuracy of measurement superior to the vernier caliper is obtained by means of a finely threaded screw, e.g. 16 threads per cm, to which the graduated 'thimble' is attached. The required reading is a combination of the main scale and the thimble scale. Increased precision is obtained by increasing the thimble diameter to, say, 60 mm, thus giving an enlarged scale, and in some cases a circular vernier scale is incorporated. In the inset to Fig. 5.4 a sample setting is given and the reader should convince himself that the required dimension is again 20.3 mm. If in good condition, a standard 25 mm micrometer should be capable of reading to within 0.005 mm. Typical sources of error are zero error, wear and tear, dust particles between mating faces, differential temperature effects and reading errors due to inexperience or carelessness.

The *toolmaker's microscope* is a useful tool which incorporates two micrometer heads attached to a floating surface table on which the component to be measured is located. Sizes can thus be obtained in two dimensions and the microscope attachment ensures consistent alignment, apart from the increased accuracy due to enlargement of the component.

Mention should be made of the use of *slip gauges* in the calibration of micrometers and vernier scales. These are small steel blocks with two highly

Fig. 5.3. Vernier calibration: (*a*) nomenclature; (*b*) sample setting.

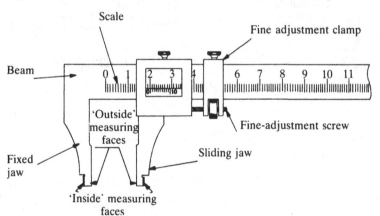

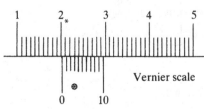

* + ⊛ =20.3 mm

finished parallel surfaces such that when two blocks are brought together with a sliding action they will stick together due to molecular attraction. The individual slip gauges are of varying thickness, so chosen that, for example, from a set of 87 blocks, any dimension in the range 0.500–100.000 mm can be obtained to within an accuracy of 0.001 mm. Micrometers and verniers can be checked for accuracy against a range of slip gauge stacks, the accuracy of which is dependent on the number of slip gauges involved.

Fig. 5.4. 25 mm micrometer: (*a*) nomenclature; (*b*) sample setting.

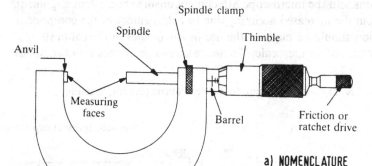

a) NOMENCLATURE

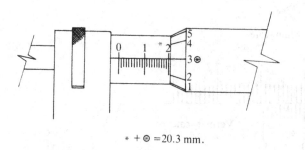

* + ⊚ =20.3 mm.

b) SAMPLE SETTING

A recent and very useful device for measuring thickness is the *digital ultrasonic thickness gauge*. This offers the significant advantage that thickness can be measured from one surface of a component. For example, the thickness of a long cylindrical pipe would be difficult to measure using more conventional methods, but is straightforward using the ultrasonic technique. The equipment operates on a pulse–echo principle, similar to that used for both sonar and radar. Very short pulses of megahertz frequency ultrasonic energy are introduced into the material to be gauged via a probe. The ultrasonic pulses travel through the component and reflect back from the opposite surface, the time for the echo to return being a function of the thickness for a given material. Preliminary calibration allows a direct digital readout of the component thickness. A variation of this device is commonly used for detecting internal flaws due to welding, casting or other manufacturing processes. (See Drury (Ref. [11]).)

5.5.2 MEASUREMENT OF TIME AND SPEED

Measurement of time becomes necessary when the frequency or speed (linear or rotational) of engineering components is required. The choice of equipment will depend on the experimental set-up and desired accuracy.

For many laboratory exercises, the use of manually operated *stopclocks* will give sufficient accuracy, although, since a ratchet mechanism is usually involved giving discrete steps of, say, $\frac{1}{5}$ s, they should be used with caution for short periods of time. Frequent checking of laboratory stopclocks, ideally before every experiment, is advisable since the rough treatment to which they are inevitably subjected has been known to cause gross errors. A useful standard for calibration is the telephone 'speaking clock' which gives a continuous 10 s repeating signal to a precision of around $\pm\frac{1}{10}$ s.

It is often the case that the variable of interest is monitored by means of some form of *transducer*, i.e. a device by which parameters such as pressure, displacement, torque, temperature, are converted to electrical outputs, in which case, the times involved in any particular transient can be measured in a number of different ways. The *cathode ray oscilloscope* (cro) is a useful general-purpose instrument for the display of a single variable. The display tube is usually about 10×8 cm in size and the time base range is normally of the order of 1 s cm^{-1}–0.5 μs cm^{-1}. It has a wide frequency response, ranging from dc to 3 MHz for a typical laboratory unit.

For permanent recording of transients use can be made of one of the large range of variable speed *graphical recorders* now available. Pen or galvanometer recorders can be used for frequencies up to about 100 Hz. For frequencies in excess of this, the *ultra-violet* recorder is particularly suitable. These are normally multi-channel instruments, the output being displayed on photosensitive charts of the order of 15–30 cm in width. Chart speeds vary typically within the range 1 mm s^{-1}–2 m s^{-1}. With a suitable choice of galvanometer element, frequencies of up to 5 kHz can be recorded. Transients can also be recorded permanently by means of *trace-recording cameras* used in conjunction with cathode ray oscilloscopes. The potential advantages of this system over graphical recorders are the higher range of frequencies, and the fact that transients can be examined prior to recording which can reduce the cost of recording material. *Storage oscilloscopes* have a facility for retaining the display semi-permanently, and in some cases a 'split-screen' facility is available for comparing the display with some standard trace.

In the field of vibrations, the *frequency counter* is possibly the most convenient and widely employed instrument for the accurate measurement of frequency and time interval. These instruments are capable of time interval measurements in the range $1 \text{ } \mu\text{s}$–10^4 s (2.8 h). They rely for their accuracy on the maintenance of a set frequency – usually in the form of ultrastable internal oscillators and, with the aid of suitable transducers, frequencies in the range 10 Hz–1.2 MHz can be monitored directly by a typical laboratory instrument.

In the measurement of rotational speed, the simplest method is to count the revolutions over a period of time by means of a stopwatch, together with a *revolution counter* connected to the rotating component. Alternatively a mechanically operated *tachometer* graduated to read directly in revolutions per minute may be used. Electric tachometers are also available which generate an output voltage proportional to speed, and, with a suitable choice of scale, can again read directly in revolutions per minute.

It is often the case that oscillatory or rotational speeds are required without interference from mechanically attached transducers. A common technique here is to use a *stroboscope*. This is basically an inert gas-discharge lamp which can be flashed intermittently at continuously variable frequencies. When the frequency of the illumination coincides with the frequency of motion, the moving part will appear to be stationary.

More recent developments in this sphere are electrically operated tachometers and counters. These can involve electro-magnetic, capacitive or photoelectric probes. To take the first as an example: irregularities in the mechanism such as gear teeth, keyways, etc., produce a momentary change in the circuit reluctance as they pass the probe, thus inducing an emf in the transducer coil, the frequency of which is proportional to the frequency of the motion, or the shaft speed, as the case may be. The output from these transducers can be arranged as a digital display using the frequency counter mentioned above.

5.5.3 MEASUREMENT OF DISPLACEMENT

(i) *Mechanical methods.* Much of what has been said on the subject of length measurement will apply equally well to the measurement of displacement. For example, the increase in length of a section of wire under load could be measured satisfactorily by means of vernier scales. Differences in manometer fluid levels can be measured accurately using a micrometer head. (See 'measurement of pressure', p. 105.) However, for the measurement of static or quasi-static displacements use is normally made of *dial gauges*, which can be adapted to suit most circumstances and give moderate precision. The basic instrument is shown in Fig. 5.5, the useful range in this case being 0–10 mm. The 50 mm diameter main dial is graduated in 100 divisions of 0.01 mm and the small dial in 10 divisions of 1 mm, each corresponding to one complete revolution of the main dial. The principle of operation is a rack and pinion mechanism with a spring to keep the measuring tip in contact with the workpiece. The main dial may be manually rotated to facilitate zero adjustment. Calibration is readily checked using a micrometer jig, or slip gauges.

Fig. 5.5. 10 mm dial gauge.

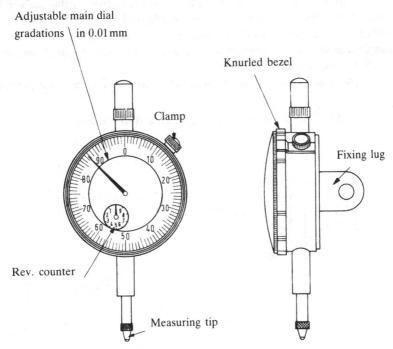

Adjustable main dial gradations \ in 0.01 mm

Knurled bezel

Clamp

Fixing lug

Rev. counter

Measuring tip

For the measurement of strain, e.g. in a tensile test, use is sometimes made of *mechanical extensometers*. There have been many variations in the design of these over the years, the most common being the Huggenberger and Johansson types. They rely on mechanical methods for the magnification of the change in the gauge length and have the advantage that they can be re-set to extend the useful range. However, on account of their physical size, cost and the need for individual reading, they are normally unsuitable for use in large numbers, in inaccessible locations, or in hostile environments. Mechanical extensometers have been largely superseded by electrically operated displacement transducers, as discussed in (iii) below.

(ii) *Optical methods*. Rotational displacement, in particular, is conveniently measured by optical methods. The advantages of using light as a means of measurement are that (a) it travels in a straight line and can act as a long lever arm to magnify rotation, and (b) the mass of light can be taken as zero and hence there is no interference with the moving part.

A typical set-up is shown in Fig. 5.6 using a *theodolite* (or a telescope fitted with cross-hairs), a small mirror and a graduated scale. The theodolite is placed at a distance l from the structure and sighted on the mirror attached to the specimen such that the scale is seen in the field of view. If the reading from the cross-hairs on the scale changes by Δs, then the angle of rotation of that part of the specimen to which the mirror is attached will be $\Delta s/(2l)$ rad. This basic principle can be adapted for use in a variety of situations. For example, with a little ingenuity it is possible, with the use of simple mechanisms, to convert linear displacement to rotational, and thus use the same optical technique.

For more accurate measurement of rotational displacement the *autocol-*

Fig. 5.6. Optical measurement of rotational displacement.

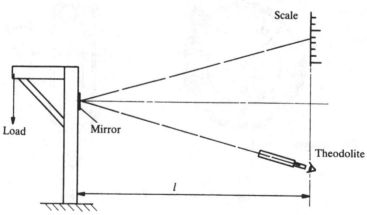

limator principle is sometimes used. This is shown schematically in Fig. 5.7 (Evans & Taylorson (Ref. [12])). It consists of a micrometer microscope *A*, an illuminated cross-wire *B*, a collimating lens *C* situated such that *B* is in the focal plane, and a high-quality reflecting surface *D*. As illustrated, an angular movement θ of the reflecting surface results in a movement of the cross-wire image by a distance *d*, which can be measured by the micrometer microscope. The required angular rotation will then be given by $\theta = d/(2f)$ rad.

(iii) *Electrical methods.* There is a variety of electrical methods now available involving *displacement transducers* of one kind or another. These have the advantage over mechanical methods in that they can be used for dynamic measurements. One common type employs varying inductance as the output to be recorded. This is produced by the movement of a permeable core in the form of a rod of the order of 3 mm diameter inside a double coil arrangement, as shown in Fig. 5.8. The coils are wound in such a way that the impedance-core displacement relationship is linear, the two coils forming two arms of a variable inductance bridge circuit. Transducers of this type are supplied over the range 0–3 mm to 0–250 mm.

The linear variable differential transformer (lvdt) is another common displacement transducer constructed in a similar way to the inductive type transducer, although with a different operating principle. In this case (Fig. 5.9) there are three coils involved. When the centre primary coil is energised with an alternating voltage supply, voltages are induced in the two secondary coils which are connected in such a way that the differential voltage output varies linearly with the displacement of the core. In both the above categories of displacement transducer the resolution is theoretically infinite, the effective resolution depending on the associated recording equipment.

Fig. 5.7. Autocollimator principle.

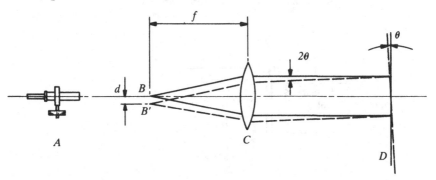

Another category of displacement transducer relies on change of capacitance between two or more movable plates, insulated from each other. These can be extremely sensitive devices but, because of difficulties in associated instrumentation, are not in common use and will not be discussed further here.

Probably the most common and certainly the most adaptable tool for use in the measurement of displacement is the *electrical resistance strain gauge*

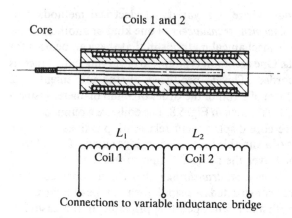

Fig. 5.8. Variable inductance displacement transducer.

Fig. 5.9. Differential transformer displacement transducer.

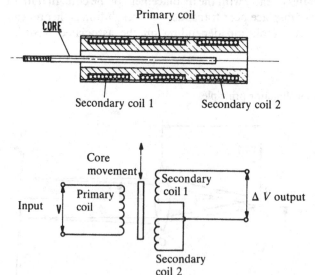

and, this being the case, it will be considered here in some detail. Emphasis will be given here to its transducer applications and later to its use in the field of stress analysis. In principle, when a length l of wire is pulled in tension, l increases and the cross-section A decreases thus increasing the resistance $\rho l/A$ of the wire, ρ being its resistivity. The strain gauge is virtually a length of wire made up in a grid form on a suitable non-conducting backing material. When the gauge is bonded to a test piece, in our case the transducer, it will be forced to strain either in tension or compression as the transducer is deformed by movement of the structure under test.

Although wire strain gauges are still used, the grid is more often constructed by an accurate etching process from foil material of the order of 0.004 mm thick. A typical foil gauge geometry is shown in Fig. 5.10. Gauge lengths vary in the range 0.4–100 mm, the most common sizes being 3 mm and 6 mm. Provided all necessary precautions are taken and the gauge is properly bonded, mechanical strain can be measured with a precision as good as $\pm 1 \times 10^{-6}$.

Change in resistance of the strain gauge is usually measured by use of equipment based on the Wheatstone bridge circuit; this is shown in its basic

Fig. 5.10. Foil strain gauge terminology (courtesy of Welwyn Strain Measurement Ltd).

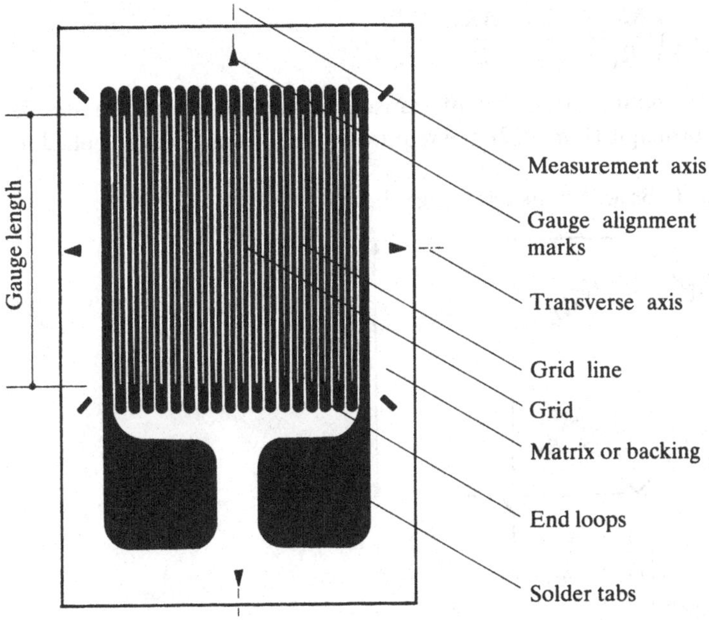

Gauge length

Measurement axis

Gauge alignment marks

Transverse axis

Grid line

Grid

Matrix or backing

End loops

Solder tabs

form in Fig. 5.11. The transducer strain gauge is R_1, and R_2, R_3 and R_4 are either additional strain gauges or stable resistors, depending on the technique employed. The bridge can be used as a null-balance system in which case the change in resistance of the active gauge, R_1, causing unbalance of the bridge, can be obtained by adjusting the resistance R_4 by ΔR_4 such that balance is restored (i.e. $\Delta V = 0$). From basic Wheatstone bridge theory we have the 'ohmic' strain

$$\frac{\Delta R_1}{R_1} = \frac{\Delta R_4}{R_4}.$$

The mechanical strain is given by

$$\frac{\Delta R_1 / R_1}{K},$$

where K is the gauge factor provided by the strain gauge manufacturer. However, in transducer applications, there is no need to use K since we are interested only in obtaining an output proportional to displacement, which can be subsequently calibrated.

As a direct read-out device the bridge can be used in a variety of ways. Consider the most general case where all four arms of the bridge are active strain gauges attached to the transducer. It can be shown (Dally & Riley (Ref. [8])) that the bridge output will be

$$\Delta V = \frac{V}{4} \left(\frac{\Delta R_1}{R_1} - \frac{\Delta R_2}{R_2} + \frac{\Delta R_3}{R_3} - \frac{\Delta R_4}{R_4} \right). \tag{5.2}$$

Many strain gauge displacement transducers are based on the simple cantilever principle (Fig. 5.12), the required displacement being applied to

Fig. 5.11. Basic Wheatstone bridge circuit.

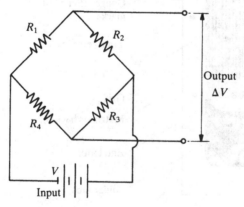

the free end of the cantilever. With one active gauge, R_1 say, and stable resistors in the other three bridge arms, the transducer would have an output of

$$\Delta V = \frac{V}{4} \cdot \frac{\Delta R_1}{R_1}.$$

However, the sensitivity of the bridge can be increased fourfold by using four active gauges located as shown in the illustration, or alternatively it can be increased by a factor of two by using active gauges R_1 and R_3 only.

The versatility of the electrical resistance strain gauge is its chief asset in transducer applications. As an example, quite recently some important recordings were made of the vibration of ducting in a nuclear reactor plant, using crude but effective displacement transducers made from ordinary hacksaw blades, strain gauged as in Fig. 5.12, and fixed to a rigid support adjacent to the ducting in question. As in other types of measurement the ability to improvise is desirable, and with a little initiative a transducer can usually be devised to provide the required signal. Other factors, such as space limitations, interference effects of the transducer on the deflecting structure, variations in support conditions, and so on, may have to be taken into account.

The range of electrically operated displacement devices is increasing continually and it would be unhelpful to extend the discussion in the present context. However, the section would be incomplete without some mention of a most versatile transducer – the *accelerometer*. There are a number of different designs of accelerometer available, but probably the most common is the piezoelectric compression type. This depends on the principle that when an asymmetrical crystal lattice is compressed an electric

Fig. 5.12. Cantilever displacement transducer.

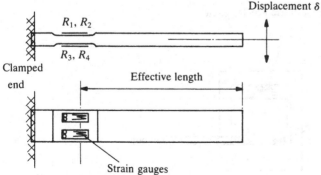

charge is produced which is proportional to the applied force. Piezoelectric elements are usually made from barium titanate or lead zirconate.

The accelerometer is shown schematically in Fig. 5.13. The piezoelectric discs are compressed by the mass, m, which is preloaded by a spring, S, and the assembly is mounted in a metal container as shown. When the accelerometer is subjected to a vibration, the piezoelectric discs will be subjected to a variable force proportional to the acceleration of the mass m. Due to the piezoelectric effect a variable potential will be developed across the output terminals, which will be proportional to the acceleration to which the transducer is being subjected. This variable potential can be monitored on an oscilloscope or on one of the recorders discussed in the previous section. By use of electrical integrating circuits the accelerometer can be used to measure velocity and displacement in addition to the determination of waveform and frequency.

One of the principal advantages of this type of transducer is that it can be made so small as to have negligible influence on the vibrating component. A typical size would be 20×16 mm diameter with a total mass of around 0.02 kg. The frequency range would be typically 2 Hz–10 kHz.

5.5.4 MEASUREMENT OF MASS, FORCE AND TORQUE

The mass of a given body is an invariant, whereas its weight depends on the local gravitational acceleration. Hence the mass of one body is obtained by comparing its weight with that of another body of known mass. This is done by the *mass balance method*, the basic principles of which are universally known. Balances range from the basic equal-arm types to those involving complex lever systems to magnify the out of balance for greater accuracy. For extreme accuracy use can be made of the highly sophisticated balance system developed by the National Physical Laboratory which can make

Fig. 5.13. Piezoelectric accelerometer.

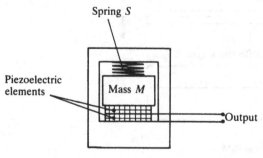

mass measurements of 'kilogrammes with microgramme accuracy' (Ref. [1]).

Clearly, the accuracy of mass balance techniques is directly dependent on the condition of the standard masses used. These should be frequently checked for, (a) excessive corrosion, atmospheric or otherwise, and (b) excessive wear due to contact with other bodies. Many of the more accurate small balances now available have the balance weights permanently housed in a sealed unit to minimise corrosion, with external controls to obviate the problem of wear.

Spring balances are commonly used to measure mass although, strictly speaking, they are only capable of measuring weight. To be frivolous, one would receive around 0.5% less value by purchasing, say, 1 kg of apples at the Dead Sea as against the summit of Mount Everest if a spring balance system were used in the purchase.

In the measurement of *force* use is commonly made of linear elastic materials in the construction of the measuring device. The force is determined by applying it to a precalibrated elastic element and measuring the resulting displacement. Household scales are an example of this principle, the elastic displacement in this case being converted to a needle rotation for convenience. Probably the most common technical example is the *proving ring* which is basically a high-tensile steel circular ring with loading blocks situated diametrically opposite each other and designed such that tensile or compressive loads can be applied. The change in diameter of the ring gives a measure of the magnitude of the applied load. The diameter change can be measured using a dial gauge, a micrometer head attachment or some form of displacement transducer, e.g. the lvdt type discussed in the previous section.

Strain gauge-type *load cells* also rely on the principle of elastic deformation and, as in the case of displacement measurement, the versatility of the strain gauge allows a broad scope in the type of load cell employed. Table 5.1 indicates the recommended gauge locations for various load cell functions, together with the relevant bridge connections for maximum sensitivity. In each case the gauges are connected such that only the force category required is monitored, e.g. for the measurement of direct loads, equal and opposite bending strains will cancel when connected as shown, and will therefore not affect the bridge balance. Although only simple tensile and bending members are illustrated in the table, it should be appreciated that more complex load cells, or *dynamometers*, as they are sometimes called, can be designed with the strain gauges situated so as to give various combinations of horizontal and vertical forces, bending moments and torques simultaneously. Octagonal or semi-circular ring elements are examples that can be conveniently used in this way.

Table 5.1

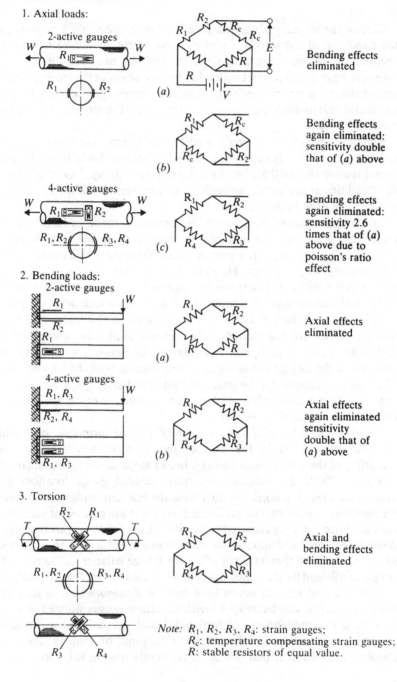

1. Axial loads:

2-active gauges

Bending effects eliminated *(a)*

Bending effects again eliminated: sensitivity double that of *(a)* above *(b)*

4-active gauges

Bending effects again eliminated: sensitivity 2.6 times that of *(a)* above due to poisson's ratio effect *(c)*

2. Bending loads:

2-active gauges

Axial effects eliminated *(a)*

4-active gauges

Axial effects again eliminated sensitivity double that of *(a)* above *(b)*

3. Torsion

Axial and bending effects eliminated

Note: R_1, R_2, R_3, R_4: strain gauges;
R_c: temperature compensating strain gauges;
R: stable resistors of equal value.

A convenient device for both applying and measuring loads simultaneously is the hydraulic (or pneumatic) *ram*. In this case the product of effective cross-sectional area and ram pressure gives the applied load. Ram cross-section areas are quoted by manufacturers to within close tolerances and hence the technique can be quite accurate depending on the method of pressure measurement used. The simplicity of the technique, however, can tempt the user to accept the resulting load figures somewhat blindly. Even with careful calibration prior to the application of the test load to account for the friction effects of the hydraulic seals, serious errors can be introduced due to increased friction resulting from unknown side loads. The possible existence of these side loads should be appreciated. In many cases means will be available for eliminating or at least minimising them.

Another hydraulic load-measuring device is the *pressure capsule*. The design might involve a flexible diaphragm or a corrugated bellows as part of a closed cell arrangement. Application of load causes a build-up in pressure which can be calibrated to give the desired output. The advantage of this type is that there are no sliding parts and hence friction effects are minimised.

Mention has been made previously of the piezoelectric accelerometer. The same principle is involved in the *piezoelectric force transducer*, the variable potential being developed in this case by the application of the force to be measured rather than the internal mass–acceleration effect of the accelerometer. Piezoelectric force transducers are suitable for dynamic and short-term static applications and a typical range of load capacity would be 15 N–1 kN.

Measurement of the power transmitted by a rotating shaft, e.g. from an experimental turbine, or an internal combustion engine, is normally achieved in the laboratory by means of some form of *prony brake*. The power produced is absorbed by a friction device running on a pulley attached to the main drive shaft. The principle is illustrated in Fig. 5.14, which shows the block type prony brake as used for high-speed shafts. The force between the wood blocks and the pulley can be varied as necessary, and by measuring the force F and the lever arm x, the output torque can be determined, from which the power can be deduced. Various designs of prony brake exist, incorporating, for example, belt friction arrangements, different lever systems and in some cases air or water cooling to absorb excessive heat.

The most common form of *torsion dynamometer* is the electric resistance strain gauge type which uses the gauge system indicated in Table 5.1 for torque measurement. Four gauges are attached to the shaft at 45° to the axis, and by connecting them as shown the torque produced can be determined, with any axial and bending strains automatically eliminated. It is necessary, of course, to use a slip-ring arrangement for the bridge connections from the rotating shaft, and this can introduce errors if sufficient precautions are not taken.

No details have been given so far on the problem of the dynamic response of the various load-measuring devices discussed. This can be complex and a detailed treatment is not within the scope of the present text. However, the existence of the problem must be acknowledged and reference can be made where necessary to more advanced textbooks (e.g. Dalley *et al.* (Ref. [9]), or Doebelin (Ref. [10])).

5.5.5　MEASUREMENT OF STRESS

Strictly speaking, stress is not a directly measurable quantity. For a simple tensile member stress can be deduced from measurements of the applied load and the cross-sectional area of the specimen. In the general case, however, stress varies considerably throughout structures and, since the strength of structures is directly related to stress level, experimental techniques have had to be devised for determining stress magnitudes, particularly at critical peak stress locations. The various methods employed in experimental stress analysis are mentioned below.

(i) *Resistance strain gauges.* Strain gauges have already been discussed in the context of transducer applications and the same basic techniques apply when they are used for experimental stress analysis. If there is sufficient symmetry in the structure and the applied loading, then right-angled pairs of gauges can be attached with their axes aligned with the anticipated principal stress directions. The two gauges of a pair can be individual

Fig. 5.14. Block-type prony brake.

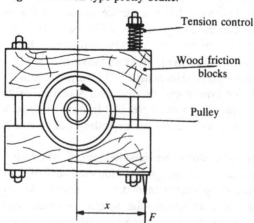

gauges, but it is now more common to use pairs of gauges that are mounted on a single backing strip. An example of such a 'rectangular rosette' is shown in Fig. 5.15. The required principal stresses can be determined from the two strain gauge readings ε_1 and ε_2 as

$$\sigma_1 = \frac{E}{1-v^2}\left[\varepsilon_1 + v\varepsilon_2\right],$$

$$\sigma_2 = \frac{E}{1-v^2}\left[\varepsilon_2 + v\varepsilon_1\right],$$

(5.3)

where E and v are the modulus of elasticity and Poisson's ratio for the material to which the gauges are attached.

Where the directions of the principal stresses are unknown it becomes necessary to determine surface strains in three directions. Any three directions will do, but it is common to attach the gauges in the form of a '45° rosette'. Again these can be three single gauges or, more commonly, will be in the form of a single three-gauge unit supplied by the manufacturer. For the 45° rosette the three strains ε_A, ε_B and ε_C at 0°, 45° and 90° to a chosen datum give principal strains from the following relationships:

$$\varepsilon_{1,2} = \tfrac{1}{2}(\varepsilon_A + \varepsilon_B) \pm \tfrac{1}{2}\sqrt{\left[(\varepsilon_A - \varepsilon_C)^2 + (2\varepsilon_B - \varepsilon_A - \varepsilon_C)^2\right]}.$$ (5.4)

The principal stresses σ_1 and σ_2 can then be obtained from equation (5.3) above.

In order to ensure that the maximum stress is determined, it is often the case that many strain gauges have to be applied to the structure under test. Individual reading of the strain gauges would be inefficient and time

Fig. 5.15. Rectangular strain gauge rosette (courtesy of Welwyn Strain Measurement Ltd).

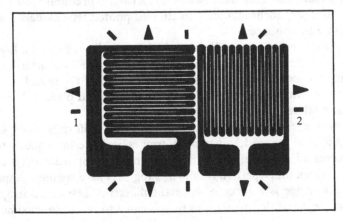

consuming, and the normal procedure would be to use a *data-logging* system. The data logger presently used in the author's laboratory has a capacity for 300 gauges and can record at a range of 25 channels per second. The data logger is also linked to a mini-computer which allows automation of the data processing and can give principal stresses directly, or even plot stress distributions if required. For further details on the use of strain gauges in stress analysis reference can be made to Refs. [8] and [23].

(ii) *Brittle coatings.* In this stress-analysis technique the component under test is coated – normally using an aerosol spray can – with a lacquer which becomes brittle on drying. As the component is loaded the coating cracks, first at high stress regions, and then more generally as the load is increased. The cracks thus indicate

(a)   the region of maximum strain; and
(b)   the direction of the maximum principal strain.

The quality of the coating and associated tools has improved considerably in recent years and it is now possible, by using a calibration procedure, to also obtain a quantitative estimate of the stress magnitude.

It is common practice to use this technique to determine the region and directions of high stress and then to attach strain gauges at these locations to give a more precise assessment of stress magnitudes. The recommended procedures for using brittle coatings are given in Refs. [2] and [3].

(iii) *Photoelasticity.* Photoelasticity is a versatile method of stress analysis that can be used both qualitatively and quantitatively. Basically the technique involves viewing a thin transparent model of a loaded structural member in a field of polarised light and then interpreting the resulting optical pattern to determine the stresses in the model. The technique relies on the property termed 'birefringence' that occurs in many transparent materials. This manifests itself in a pattern of coloured bands – termed 'fringes' or 'isochromatics' – in stressed photoelastic models when observed in a polariscope.

A *transmission polariscope* is illustrated schematically in Fig. 5.16. Light from the source first passes through a plane polariser and a quarter-wave plate which induces circular polarisation, then through the model and a second quarter-wave plate, and finally through a second plane polariser, termed the analyser, and thence to the observer.

When unstressed the transparent model will be uniformly clear. On applying a load to the model a pattern of fringes begins to form and as the load is increased the number of fringes increases proportionately. An example of such a fringe pattern is shown in Fig. 5.17 for a spanner-shaped model. Each fringe is a locus of constant difference between principal stresses $(\sigma_1 - \sigma_2)$ and this quantity in turn is equal to twice the maximum

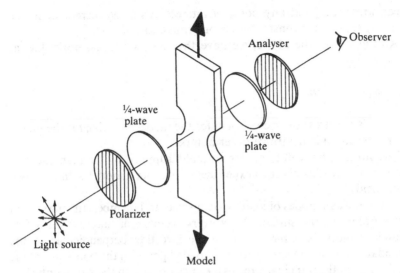

Fig. 5.16. Transmission polariscope for photoelastic stress analysis.

Fig. 5.17. Example of photoelastic fringe pattern (courtesy of Stress Engineering Services).

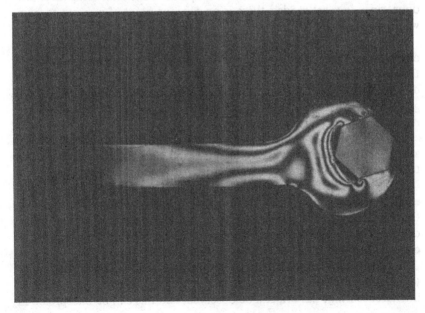

shear stress $(2\tau_{max})$ at any point of interest. Also, the increment in the quantity $(\sigma_1 - \sigma_2)$ is constant from one fringe to the next.

Arising out of the above, we have the so-called 'stress-optic' law as follows:

$$\sigma_1 - \sigma_2 = \frac{f}{t}\, n,$$

where f is the stress-optic coefficient for the model material; t is the model thickness; and n is the fringe order number.

The stress-optic coefficient f is normally supplied by the manufacturer but it can be readily checked experimentally using a model with a known stress field.

For almost all modes of loading on a structural member, the maximum stress occurs at the surface. Now every element of material on a free boundary must have zero stress in the direction perpendicular to the boundary surface. Thus, at every non-loaded point on the boundary of the model, one of the principal stresses (i.e. σ_2) is zero, and the stress-optic law reduces to

$$\sigma_1 = \frac{f}{t}\, n.$$

From this simple relationship we need only identify the order number of the fringe n to obtain the principal stress at points of interest. This is readily done, since in most models there is an obvious point of zero stress from which location the fringe order can be determined.

The photoelastic method of stress analysis has been extensively developed in recent years, and many refinements exist, not least of which is the stress-freezing technique, which allows the analysis of complex three-dimensional components, and the photoelastic coating technique, which gives the facility to test real structural members. Refs. [2] and [8] are recommended for further reading.

(iv) *Other optical techniques.* Two further optical techniques used in stress analysis are the Moiré fringe method and the thermal emission technique and these are mentioned briefly to round off this section.

The Moiré effect occurs when two similar but not quite identical grids of equally spaced lines are arranged so that one grid can be viewed through the other. The effect can be seen from public highways when approaching overhead bridges with fences on either side. The Moiré fringe effect can often be seen clearly as the bridge is approached.

In stress-analysis applications one grid is attached to the structure under test and this is viewed through a master grid. As the structure is loaded an interference fringe pattern can be observed, and this can be interpreted in terms of component strains and hence stresses can be assessed. Typical grid spacings for Moiré work would be in the range 20–400 lines per cm. Dally & Riley (Ref. [8]) give some details on the application and interpretation of the technique.

The stress analysis by thermal emission (SPATE) technique is a recent and exciting development. It is based on the measurement of the thermoelastic effect in materials when the imposition of stress by dynamic loading produces minute temperature changes that are stress related. The resulting infrared emission caused in the test component can be recorded as quantitative colour contour maps and the resolution is claimed to be as good as $1 \, \text{N} \, \text{mm}^{-2}$. The equipment can be used for a wide range of materials including plastics and all metals. A description of the SPATE technique is given by Mountain & Webber (Ref. [17]).

5.5.6 MEASUREMENT OF PRESSURE

In a fluid system *pressure* means the force acting upon and normal to unit area of the system. An *absolute pressure* refers to the absolute force acting normally on a unit area of the fluid system. *Gauge pressure* represents the difference between the absolute pressure and the local barometric pressure. Very often, in practice, the readings obtained from pressure gauges are relative to the barometric pressure. It is important that the distinction is clearly understood as serious errors in results can sometimes be traced to this point. '*Vacuum*' is the amount by which the barometric pressure exceeds the absolute pressure (i.e. it is then a negative gauge pressure).

It may surprise the reader to learn that it is possible for a liquid to sustain a negative absolute pressure (i.e. a tensile stress)! Osborne Reynolds (Ref. [20]) demonstrated before a learned society how he had managed to support a column of mercury to more than twice the height of the barometer by the internal cohesion of the liquid. The important factor was the great care he took in removing the dissolved gases before inverting the mercury.

More recently (and also more easily accessible to interested students) Ryley (Ref. [22]) devised a simple centrifugal apparatus to test the tensile strength of both raw tap water and boiled, de-aerated filtered water. He and his team found values of stress at rupture in excess of 2.6 bar for boiled, de-aerated water.

In the laboratory, students learn to use a variety of pressure-measuring instruments. Because of its extreme simplicity and reliability, the liquid-filled manometer is a popular instrument for the measurement of steady fluid pressures. Many versions of the manometer principle are in use and space permits discussion of only a few of these.

The *U-tube manometer* (Fig. 5.18(*a*)), is the simplest version and consists of a constant bore glass tube in the form of a long U, partly filled with suitable liquid such as water, alcohol or mercury. To measure the pressure of a fluid which is less dense and immiscible with the manometer fluid, a connection is made to one limb of the U-tube while a suitable reference pressure is applied to the other limb. The vertical displacement, z, of the manometer fluid gives an indication of the applied pressure difference, Δp. If ρ_m is the density of the manometric fluid, ρ_f that of the fluid whose pressure is to be measured and g the gravitational acceleration, then

$$\Delta p = (\rho_m - \rho_f)gz.$$

When a low-density gas (e.g. air at ambient pressure) is displacing the manometric fluid level, $\rho_f \ll \rho_m$, and it is then acceptable to use the approximation

$$\Delta p = \rho_m gz.$$

The U-tube manometer needs no calibration except for that of length, the relative density of the manometric fluid and to a limited extent the local value of g. To realise the high inherent accuracy of manometers a number of corrections must also be applied. For accurate measurements corrections for the thermal expansion of the length scale and for the fluid density would be applied. Additional error can be caused by difficulty over reading z due to the meniscus formed by surface tension. This source of error can become pronounced if the inside surfaces of the tubes are 'dirty' (e.g. mercury manometers are particularly prone to this problem due to the oxidation of mercury) or their bores are non-uniform. Surface tension effects are greater for tubing of fine bore and alcohol is often preferred to water as its surface tension is only about one-third that of water. Note that the surface tension of water can be reduced significantly by adding a small proportion of some wetting agent. One drawback in using alcohol (in manometers anyway!) is the gradual change in relative density because of water absorption. Alcohol is hygroscopic and its density increases if it is in contact with gases containing any water. Occasional checking of the relative density of alcohol used in manometers with a *densitometer* is good engineering practice and lends credibility to the results obtained. A densitometer (density measuring device) essentially weighs a known volume of liquid.

Pressures less than about 20 mm water are difficult to measure on a vertical U-tube manometer to an accuracy better than ± 0.5 mm. Sensitivity can be increased by a factor of up to 10 times by inclining the tubes from the vertical. However, angles approaching 5° from the horizontal are not recommended unless the tubes used are very straight. The bore should be about 3 mm for low angles of inclination and fluid of low surface tension used to preserve a satisfactory meniscus shape. Fig. 5.18(*b*) shows an adaptation of the simple U-tube which has a very wide internal diameter in one limb. Readings need only be taken from the narrow limb since the fluid level in the reservoir limb remains substantially constant. A correction for differences in surface tension effects may be necessary because of the two different tube diameters.

Compared with the U-tube manometer, the *micromanometer* (Fig. 5.19) can improve measurement accuracy by an order of magnitude but at the cost of some loss of convenience. The reservoir can be adjusted in height by means of a micrometer screw, which changes the level of the liquid in the inclined tube. The inclined tube is marked at one point, which is the *datum*.

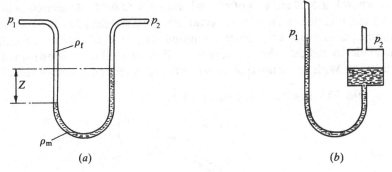

(*a*)   (*b*)

Fig. 5.18. Two versions of the U-tube manometer.

Fig. 5.19. A micromanometer.

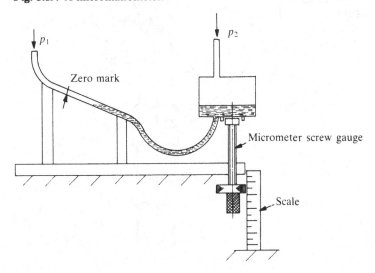

With an applied pressure difference, $p_1 - p_2$, the reservoir height is adjusted until the level of liquid in the inclined tube is returned to the datum mark. The change in height can be accurately found on the micrometer gauge.

The improved accuracy of this instrument is due to the fact that the meniscus position remains the same for all pressure difference readings. Thus, surface tension effects due to tube non-uniformity or to dirt are reduced. A low-power magnifier is often used to observe the meniscus. The accuracy of the micromanometer can be as good as ± 0.002 mm of manometer fluid.

For measuring pressures outside the range of the mercury-filled manometer, the *Bourdon tube*, because of its simplicity and versatility, is the basis of many types of pressure gauge. In its simplest form the Bourdon tube shown in Fig. 5.20 consists of an oval section tube bent into a circular arc. One end is sealed and free to move; the other end is rigidly fixed and is open for the transmission of pressure. By means of a system of linkages a pointer is made to indicate the pressure applied to the tube, as shown in the diagram. With some internal pressure the tube section becomes rounder

Fig. 5.20. Bourdon tube pressure gauge.

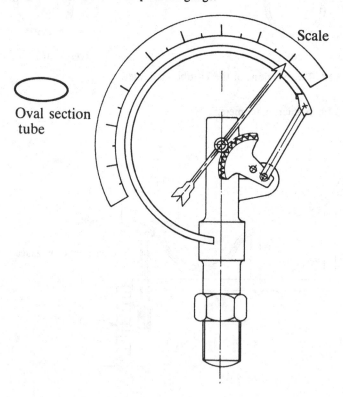

Scale

Oval section
tube

and this causes it to become straighter. It is this motion, amplified mechanically, which is indicated by the gauge needle.

Pressure gauges are available for the measurement of absolute, 'gauge' or differential pressures over wide ranges. It is essential that periodic checks of the calibration are made to offset the effects of wear on moving parts and ageing of elastic components.

Pressure transducers which convert pressure levels into electrical signals are finding frequent application and can take many forms. Pressure is applied to an elastic element such as a Bourdon tube or a diaphragm or a bellows unit and the resulting movement is measured by some form of displacement transducer as discussed previously, (iii) 'Electric methods', p. 91. In common with other forms of transducer, the pressure transducer lends itself to the measurement of fluctuating pressures. Many different techniques are in use, for example:

(a) resistance changes of strain gauges bonded to a diaphragm;
(b) diaphragm movement as measured by an inductive-type displacement transducer or linear variable differential transformer (lvdt);
(c) capacitance changes produced by the motion of a diaphragm forming one plate of a capacitor. This type is used for measuring rapidly fluctuating pressures which occur in internal combustion engines, for example;
(d) resistance changes by the wiper action of a potentiometer.

The calibration of many kinds of pressure measuring devices, e.g. Bourdon tubes, piezoelectric pressure transducers, engine indicators, etc., is usually performed with a precision instrument called the *dead-weight tester*. Fig. 5.21 shows schematically a dead-weight calibration unit in which the hydraulic pressure of a fluid is increased by winding in a screw attached to a piston. Various loads, W, are applied to the vertical spindle of area A. When the spindle is lifted by the hydraulic pressure, the pressure force is equal to the total gravity force of the load and spindle. The effect of friction error on the calibration can be almost eliminated by using a low viscosity oil as the hydraulic fluid and rotating the spindle while a reading is being taken so that kinetic friction applies rather than the 'stick-slip' condition of static friction.

5.5.7 MEASUREMENTS IN FLUID FLOW

In many experimental studies it is necessary to measure the magnitude and direction of the flow velocity at a point in the flow and to find how this varies across the flow. In some flow situations, e.g. inside parallel-sided ducts, the flow direction is known with reasonable accuracy and then only the velocity needs to be measured.

1. *Velocity measurements.* The velocity, V, at a point in a constant density fluid flow may be obtained most conveniently but indirectly from measurements of the *total pressure*, p_0, and the *static pressure*, p, at the same point. With the fluid density, ρ, known, the velocity can be determined from Bernoulli's equation (Gibbings, Ref. [13]), i.e.

$$p_0 - p = \tfrac{1}{2}\rho V^2.$$

This equation can also be used for low-speed gas flows with relatively small error. For air at 20 °C and 1 bar at a *Mach number M* of 0.2 (corresponding to a true airspeed of 68.36 m s^{-1}), Bernoulli's equation underestimates velocity by only about 0.5%. Mach number is defined as the flow velocity divided by the sonic velocity. For higher Mach numbers $M < 1$ the compressible flow formula must be employed, i.e.

$$\frac{p_0}{p} = \left(1 + \frac{\gamma - 1}{2} M^2\right)^{\gamma/(\gamma - 1)},$$

where γ is the ratio of the specific heats of the gas.

On the surface of any solid body immersed in a flowing fluid there is some point, usually the most forward point, at which the fluid is brought to rest. At this point the pressure is the total pressure, sometimes called the *pitot pressure* or the *stagnation pressure*. In subsonic flow ($M < 1$) this pressure is relatively easy to measure. Fig. 5.22(a) shows an instrument called a *pitot tube* which is arranged in a flow so that a small pressure hole at the end of the tube is at or very close to the stagnation point. The instrument will

Fig. 5.21. A dead-weight tester.

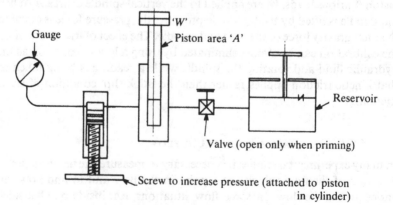

Gauge

'W'

Piston area 'A'

Reservoir

Valve (open only when priming)

Screw to increase pressure (attached to piston in cylinder)

record the total pressure without appreciable error even when not perfectly aligned with the direction of flow. For most purposes it is sufficient to sight the tube within about 5° of the flow direction using some simple device such as cotton thread streaming in the flow. Alternatively, the instrument can be orientated until a maximum reading is obtained (Bryer & Pankhurst (Ref. [7])).

By enclosing the pitot tube within an open-ended outer tube the pitot reading can be made extremely insensitive to flow direction. This can be very useful in situations where it is not possible to orientate the probe to

Fig. 5.22. (*a*) Pitot tube inside a duct measuring total pressure relative to atmospheric pressure. (*b*) Shrouded pitot tube – the shroud tube redirects a flow inclined at angle α to the tube axis so that the pitot tube receives the total pressure ($\alpha < 40°$).

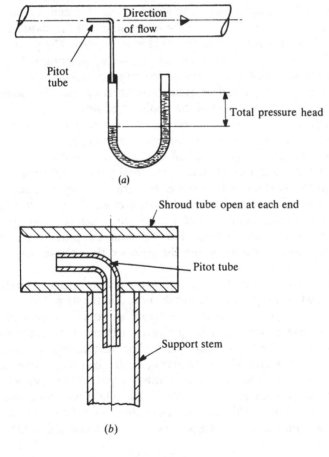

maximise the reading and the flow direction is uncertain. Fig. 5.22(*b*) shows such a *shrouded total pressure* tube which will remain insensitive to yaw over ±40°.

In supersonic flow ($M > 1$) pitot tubes do not record total pressure correctly. This is because a *detached shock wave* forms upstream of a blunt body in the supersonic stream, resulting in some loss of total pressure. The total pressure recorded by the pitot tube is that of the subsonic flow *after* the shock wave. By employing normal shock wave relations (or tables) the free-stream supersonic Mach number can be determined from the pitot tube and static pressure observations. One problem which occurs here is that of determining the local static pressure. The difficulty may be overcome with the Goodyer probe (Ref. [14]) which decelerates the supersonic flow isentropically to a subsonic speed upstream of the measurement tapping.

Although static pressure could be measured by any form of pressure probe if it moved with the fluid, this is seldom possible and a special form of stationary probe is normally used. The accurate measurement of static pressure is inherently difficult because the introduction of the stationary probe into the flow changes the pressure field in its vicinity. The pressure around a probe varies from point to point in a way largely determined by its shape and orientation to the flow direction. The pressure-sensing holes on a probe are located where the surface pressure is equal or related in some known way, to the static pressure of the undisturbed flow.

A simple type of static pressure probe consists of a body of revolution with its axis aligned to the flow direction, supported by a lateral stem at the downstream end of the body (Fig. 5.23(*a*)). This particular instrument also includes a forward-facing hole for measuring total pressure. Because the flow is brought to rest at the nose the pressure there is greater than that of the undisturbed flow. As the flow accelerates around the shoulder the pressure falls very rapidly below that of the free stream before rising again slowly towards the free stream value. Further along the probe the stem causes a further flow deceleration and the pressure rises above the free stream level.

Fig. 5.23(*b*) gives an example of the variation in pressure errors along the probe length due to the separate effects of the nose and the stem. The holes are positioned at the point where the two errors are equal and opposite. This design principle is employed in the precision instrument often referred to as the *NPL standard pitot–static tube*. The instrument is rather cumbersome to use in many flow situations but it is frequently employed as a measuring standard when calibrating other types of flow velocity measuring probes. The calibration of the NPL standard instrument was originally performed on a whirling arm. The subject of airflow measurements using NPL and other probes is covered in a wealth of detail by Ower

& Pankhurst (Ref. [18]) and it is worth supplementing this reading with Bryer & Pankhurst (Ref. [7]).

Velocity measurements can only be made with pitot–static instruments when the flow is steady, or very nearly so. The response time of such instruments is long because of the large-flow resistance of small-diameter pressure holes and connecting leads. In fluctuating flows the measurement of velocity is nearly always made with some form of *hot-wire anemometer*.

The temperature attained by an electrically heated wire immersed in a flowing fluid depends upon the velocity of flow past the wire and its orientation to the flow. For a hot wire placed normal to the flow it is possible to calibrate the wire to indicate velocity in terms of wire temperature.

Fig. 5.24 shows a simple version of a hot-wire anemometer probe which forms, in effect, one arm of a Wheatstone bridge circuit. The wire element is often made of very fine gold-plated or platinum-plated tungsten wire (0.002–0.005 mm diameter), soldered or welded to two probe supports (0.5–3 mm apart). The bridge supplies current to the wire, heating it to a *constant temperature* between 200 and 300 °C. Constant wire temperature, which can be held within very small limits, is made possible by electrical

Fig. 5.23. (*a*) NPL standard pitot–static tube. (*b*) Variation of static pressures along tube.

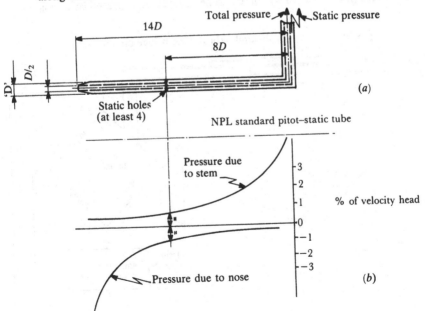

feedback from the amplifier. Thus, when the fluid velocity, V, changes, the heating current, I, to the wire also changes to maintain the bridge in balance. The hot-wire temperature, T_w (as determined by its resistance, R_w), is maintained constant. Essentially, for thermal equilibrium, the ohmic heating of the wire balances the convective heat loss from the wire, i.e.

$$I^2 R_w = hA(T_w - T_f),$$
(5.5)

where h is the coefficient of heat transfer; A is the area of heat transfer surface; and T_f is the temperature of fluid.

For a range of velocities h is mainly a function of flow velocity, i.e.

$$h = A + BV^n,$$
(5.6)

where $n \simeq 0.5$, A and B are constants for a given probe. Combining equations (5.5) and (5.6) produces the useful expression

$$I^2 = A_1 + B_1 V^n,$$
(5.7)

where A_1 and B_1 are constants obtained by calibration. The value of the index n in equation (5.7) is often assumed to be 0.5, but the true value can be determined either by a process of trial and error or more quickly with an electronic 'lineariser'.

The accurate measurement of the average flow velocity (i.e. the steady component of velocity) requires that a given hot-wire probe must be calibrated in the fluid in which it is to be used. It is calibrated by exposing it

Fig. 5.24. Hot-wire probe and Wheatstone bridge circuit.

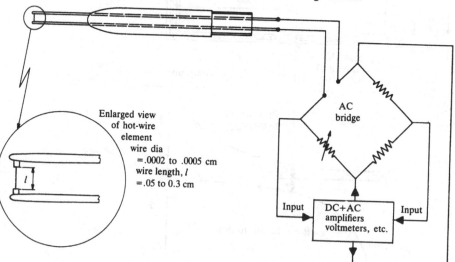

Enlarged view
of hot-wire
element
wire dia
= .0002 to .0005 cm
wire length, l
= .05 to 0.3 cm

AC
bridge

Input   DC+AC
amplifiers
voltmeters, etc.   Input

to known velocities measured, for example, with an NPL standard pitot–static tube, and its output current or voltage drop recorded for a range of velocities.

Practical problems occur with hot-wire anemometers because of the limited strength of the fine wires and calibration changes due to dirt accumulation on the wire sensor. Unless the flow is clean the calibration can change in a matter of minutes and even wire breakage is not unknown due to the impact of larger particles. The problem is much reduced with frequently changed upstream filters. At high flow velocities the wire can break due to aerodynamic loading of the wire, vibration of the wire and vibration of the supports. *Hot-film probes* overcome some of these problems as they have a more robust construction than hot-wire probes. The sensor is a thin film of nickel sputtered on to a quartz substrate, usually a narrow wedge shape, and covered with a quartz film for protection. The hot-film probe simply replaces the hot-wire probe in the anemometer bridge circuit. It has the great advantage of being able to withstand more arduous environmental conditions than the hot-wire probe. Film sensors are used for measurements in liquids with velocities up to 25 m s^{-1} and in gases with velocities up to 500 m s^{-1}. Although wire sensors can also be used in high-speed gas flows they may be used in liquids at velocities less than 5 m s^{-1} provided the liquid is non-conducting.

One very important merit of the hot-wire and hot-film anemometer is its capacity to measure the turbulence microstructure of gas and liquid flows. This is possible because of the very fast response and small size of the probes. The practical upper limit of frequency for hot-wire probes in air is stated to be about 350 kHz.

5.5.8. LASER ANEMOMETRY

An important relatively recent advance in flow-measurement technique is provided by the *laser–Doppler anemometer* (LDA). This instrument measures local flow velocity and its microstructure and thus is in competition with the hot-wire or hot-film anemometer described earlier. The main advantages of the LDA compared with other anemometers include,

(i) velocity is directly measured rather than being indirectly inferred from heat transfer considerations;

(ii) no flow interference caused by a physical object;

(iii) high spatial and temporal resolution – a cube of side 0.2 mm is not untypical and frequency response can be in the MHz range.

Disadvantages include the need for fluid and wall transparency and the high cost and complexity of the apparatus.

Fig. 5.25 shows, schematically, a *reference beam LDA* in which light from a laser is divided into an *illuminating beam* and a *reference beam*. Light from the illuminating beam is scattered by particles in the flow in the direction of the reference beam. The operation of a LDA is essentially dependent on the presence of particles to scatter light and these must have a certain size (e.g. 2–5 μm) to produce good Doppler signals but still be small enough to follow the turbulence of a fluid flow. The particles may occur naturally in the flow or else they must be introduced by 'seeding'. In air flows an atomised liquid is often used for seeding and in water flows polystyrene latex or even air bubbles are employed.

When the light from the two beams is combined in the photoamplifier an output signal is produced containing a beat frequency equal to the Doppler shift frequency, f_D, produced by the motion of the particles. This frequency is determined by the spectrum analyser and it can be shown (Dally *et al.* (Ref. [9])) that

$$f_D = \frac{2V \cos \alpha}{\lambda} \sin \frac{\theta}{2},$$

where V is the velocity of a particle; λ is the wavelength of the laser light; α is the angle between particle velocity vector and the normal to the bisector of the two beams; and θ is the angle between the two beams.

Fig. 5.25. Reference beam laser-Doppler anemometer.

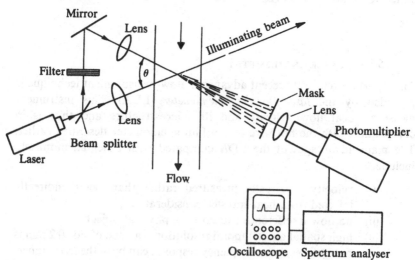

No calibration is required for the LDA. As the wavelength of the helium–neon laser is known to an accuracy of 0.01%, and as the electronics can provide very accurate determination of the Doppler-shift frequency f_D, the accuracy of velocity measurement is controlled by how accurately the angle θ is measured.

2. *Flow direction.* It is often necessary to measure the local direction of flow as well as the velocity and the usual method is to employ some form of differential pressure device. One of the simplest of these is the *yaw tube* (Fig. 5.26(a)), consisting, in a two-dimensional flow, of a circular cylinder with its axis normal to the plane of flow. Two small holes (pressure tappings), spaced at about 60° apart facing the oncoming flow, are connected to the two sides of a U-tube (or other) manometer. The tube is rotated through small angles until the difference between the pressure readings is zero, at which point the yawmeter is said to be 'balanced' or 'nulled'. Provision is usually made for the angle of flow to be read off a vernier protractor.

The same principle is employed in other, more sophisticated designs of yawmeter. Fig. 5.26(b) shows a *claw probe* and Fig. 5.26(c) a *wedge yawmeter*, both fitted with a pitot tube and called a combined probe, which are used in situations where flow direction and velocity change rapidly across a flow. These and other types of pressure-sensing yawmeter probes are dealt with in some detail by Bryer & Pankhurst (Ref. [7]). All of these combined instruments are capable of measuring total, static and dynamic

Fig. 5.26. Some flow-direction sensing probes: (*a*) cylindrical, (*b*) claw-type, (*c*) wedge-type.

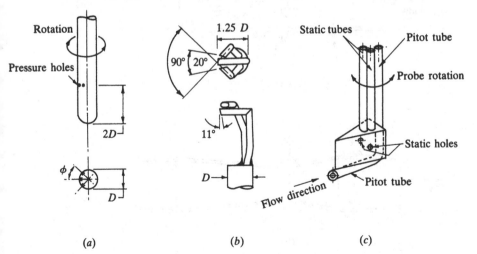

(*a*) (*b*) (*c*)

pressures as well as flow direction. It is normally necessary to calibrate these pressures by comparing readings with another instrument, usually an NPL standard pitot–static tube in a wind tunnel.

3.*Mass flow*. Although wedge yawmeters are suitable for detailed flow exploration and can, in principle, be used for measuring overall mass flow through a duct by traversing them across the duct, this would be tedious and rather impracticable. The simplest method of measuring the flow through a duct is to pass the whole flow through some sort of constriction and measure the pressure drop across it. Fig. 5.27 shows two commonly used flowmeters which are designed on this principle.

The *orifice plate flowmeter* (Fig. 5.27(*a*)) is relatively crude but inexpensive to install. It is mostly used when the irrecoverable pressure drop caused by the plate in unimportant. The *venturi tube* (Fig. 5.27(*b*)) causes little loss in total pressure because of the smooth deceleration of the flow after the contraction; however, it is more expensive.

When these flowmeters are made to standardised proportions, the pressure difference measured $(p_1 - p_2)$ can be very large and may be

Fig. 5.27. Flowmeters based on stream pressure changes: (*a*) orifice plate flowmeter, (*b*) venturi flowmeter.

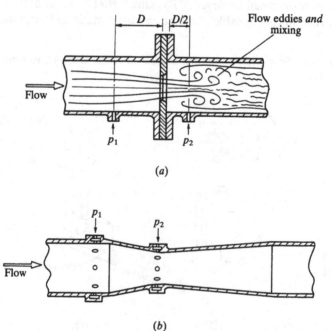

(*a*)

(*b*)

measured to an accuracy better than 99% without using a manometer of high sensitivity.

The *vortex shedding flowmeter*. When a circular cylinder is placed in a uniform stream with its axis perpendicular to the direction of flow, eddies often called vortices are shed from alternate sides of the cylinder in a regular pattern known as Karman vortex street, as shown in Fig. 5.28(a). The shedding frequency, f, the diameter of the cylinder, d, and the flow velocity, V, are related by a dimensionless number called the Strouhal number, S. Thus,

$$S = fd/V.$$

Measurements show that S is almost constant over a wide range of Reynolds number Re based on the cylinder diameter, i.e. $0.2 < S < 0.21$ for $300 < Re$ $150\,000$ where $Re = Vd/v$ and v is the kinematic viscosity of the fluid.

Fig. 5.28. Measurement of mass flow rate through a duct: (a) Karman vortex street, (b) a vortex shedding flowmeter.

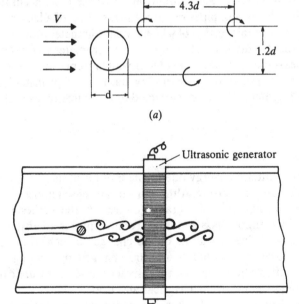

(a)

Ultrasonic generator

(b)

A recent development is the Technitron *vortex pipe flowmeter* (Fig. 5.28(b), which employs a vortex-sensing technique to measure the frequency, f, of vortices shed of a small-diameter cylinder installed across the diameter of a short pipe. An ultrasonic beam crosses the path of the vortices and each pair of vortices causes one cycle of amplitude modulation to be impressed on the carrier wave as a result of beam scattering by the oppositely rotating vortices. As the frequency is a function of flow velocity the flow rate can be directly measured. Calibration studies have shown that the results are unaffected by the temperature, density or pressure of the fluid (Dally *et al.* (Ref. [9])).

The advantages of this device are its low error (1% of full scale), good repeatability, low cost, linearity (0–10 V dc for full scale), low pressure drop and its rapid response.

The primary calibration of a flowmeter is, in general, based on steady flow through the meter to be calibrated and the subsequent measurement of the volume or mass of fluid that has passed during an accurately timed interval. Any stable and precise calibration obtained by this means would itself become a secondary standard against which other flowmeters could be more conveniently calibrated. Turbine flowmeters have been found especially suitable as secondary standards, some having accuracies as high as 99.8%. According to Ower & Pankhurst (Ref. [18]) the most accurate method of measuring the flow in pipes requires the insertion of a length of smooth-bore pipe, at a suitable section of which a small-diameter pitot tube can be traversed across the diameter of the pipe by a micrometer screw. The static pressure is measured by a minimum of four wall tappings in the same plane. Precautions are required to ensure the flow has no swirl by installing honeycomb flow straighteners at least ten pipe diameters upstream of the traverse plane.

5.5.9   MEASUREMENT OF TEMPERATURE

A substance possesses internal energy due to the motion of its molecules and this is manifested as temperature. Although we can sense temperature by touch or by experiencing the thermal radiation from hotter bodies, it is not readily defined like length, mass and time. Temperature is an abstract quantity that must be defined in terms of the changes in behaviour of substances as their levels of internal energy are changed by heating or cooling. Observable effects brought about by changes in temperature can be categorised as physical, chemical, electrical and optical.

The well-known *liquid-in-glass thermometer* which depends upon the relative expansion of the liquid with temperature increase is adaptable to a wide range of applications. The liquids most commonly used are mercury and alcohol. Mercury can be used for temperatures between −39 °C (its freezing point) to over 500 °C, the range near the upper limit requiring the use of special glass and a pressurised inert gas fill above the mercury. When a lower temperature limit is needed an alcohol-in-glass thermometer enables measurements to be taken to as low as −62 °C. Other liquids occasionally used in thermometers and their corresponding freezing points are: toluol (−90 °C), pentane (−201 °C).

Glass thermometers are designed for either total or partial immersion. Total-immersion thermometers are calibrated to read correctly when the column of liquid is completely immersed in the fluid whose temperature is being measured. Partial-immersion thermometers are calibrated to read correctly when immersed to a datum mark and with the exposed stem at a definite temperature. They are less accurate than total-immersion thermometers. However, it is often necessary to use both types of thermometer in conditions different from that at calibration and a correction should then be applied. The correction can be estimated by suspending a small auxiliary thermometer close to the emergent stem, as shown in Fig. 5.29. This auxiliary thermometer estimates the average temperature T_a of the emergent stem. A simple correction for the stem temperature can be applied (Doebelin (Ref. [10])), with the equation

$$\Delta T = KN(T_i - T_a), \tag{5.8}$$

where T_i is the indicated temperature of the thermometer, N is the length of

Fig. 5.29. Liquid-in-glass thermometer with small auxiliary thermometer. (NB: values shown on diagram relate to example in text.)

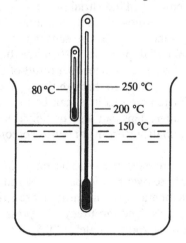

80 °C — 250 °C
— 200 °C
— 150 °C

exposed thermometric fluid in degrees; and K is the differential coefficient of expansion.

The following example illustrates the application of equation (5.8) to the typical situation arising when a total-immersion thermometer is only partially immersed in order to observe the reading.

Example. A mercury-in-glass thermometer indicated 250 °C when the column of mercury was immersed in a liquid up to the 150 °C mark. If the mean temperature of the mercury not immersed was 80 °C what was the true temperature of the liquid? The differential expansion coefficient between the mercury and glass is $1.6 \times 10^{-4}/°C$.

$$\Delta T = 1.6 \times 10^{-4} \times 100 \times (250 - 80) = 2.7 °C.$$

Thus, to a first approximation, the temperature of the liquid was 252.7 °C.

If a partial-immersion thermometer is used at the correct immersion but at an ambient temperature, T_a, different from that of the calibration air temperature, T_c, then the correction required is

$$\Delta T = KN(T_c - T_a). \tag{5.9}$$

Suppose the thermometer in the example given above is of the partial-immersion type with its datum mark at 150 °C and its calibration air temperature $T_c = 60 °C$. With the other data unchanged then, using equation (5.9),

$$\Delta T = 1.6 \times 10^{-4} \times 100 \times (60 - 80) = -0.32 °C.$$

The use of liquid-in-glass thermometers is often limited by the fragility of the glass and the difficulty of reading the scale because of inaccessibility. Also, as the reading is only visual, they cannot be used in automatic data-recording systems or in automatically controlled industrial processes. The *mercury-in-steel thermometer*, Fig. 5.30, avoids these limitations. The system comprising the bulb, capillary tube and pressure sensor is completely filled with mercury under an initial pressure. When the bulb experiences a temperature change, the pressure changes as a result of the differential volume change. The pressure transmitted through the capillary tube (which can be up to 60 m long for remote readings) can be read on pressure sensors such as a Bourdon gauge. This can also be linked to a potentiometer or to an lvdt to allow automatic recording or control of some industrial process to take place.

Pressure thermometers filled with mercury can cover the range from −39 °C to over 500 °C with linear response over a large part of the range. Details of various refinements such as temperature compensation are given in texts such as Doebelin (Ref. [10]). The accuracy of pressure thermometers under the best conditions is approximately ±0.5% of full scale. Adverse environmental conditions may increase the error substantially.

The difference in the expansion coefficients of two metals can be used to indicate changes in temperature. The two metals are welded or brazed together along the surfaces. Typical metals used are invar and brass. When heated, the brass expands faster than the invar, causing the strip to deflect. Long *bimetal strips* wound into the form of a helix can be used as the temperature-sensing element of thermometric devices. Bimetal elements in the form of cantilever beams, spirals and washers are cheap and deform significantly for relatively small temperature changes. They are used in thermostats to control temperature by switching the heating on and off. In electrical equipment they can act as overload switches activated by excessive current flow. They are often used in conjunction with potentiometers or lvdt's to provide a temperature-indicating instrument. In applications where accuracy is important, good-quality bimetal thermometers are available with accuracies guaranteed to about $\pm 1\%$ of full scale.

Resistance thermometers comprise a 'sensor' whose electrical resistance varies with temperature, a means of measuring resistance, and a relation between resistance and temperature. Typically, a resistance thermometer sensor consists of a fine platinum wire wound on to a mica frame and enclosed in a protective tube. The leads are taken to some form of bridge circuit by which changes in resistance with temperature are measured with high accuracy. Features that characterise resistance thermometers are stability of sensors, sensitivity of measurements and simplicity of the circuitry.

Fig. 5.30. Mercury-in-steel thermometer.

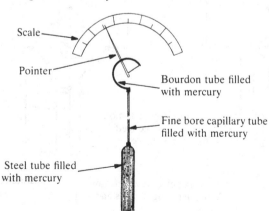

Fig. 5.31 shows resistance thermometer sensors as part of Wheatstone bridge circuits suitable for normal engineering laboratory use. For high precision the conventional Wheatstone bridge is not very satisfactory since:

(a) slide-wire contact resistance is a variable dependent upon condition of slide;

(b) temperature gradients along extension wires to sensor cause variable resistance of wires; and

(c) supply current itself causes variable I^2R heating due to changes in sensor resistance.

More advanced texts giving details of various refinements to circuits to alleviate these difficulties are Benedict (Ref. [5]) and Doebelin (Ref. [10]).

In 1821, Thomas Seebeck discovered that when the junctions of two dissimilar metals comprising a closed circuit were exposed to a temperature difference, a net emf was produced which induced a continuous electric current around the circuit. The phenomenon is called the *Seebeck effect* and involves the conversion of thermal energy into electrical energy. The subject of *thermoelectricity* which deals with the theory of the Seebeck effect and its application to the production of useful electric power is discussed concisely by Rogers & Mayhew (Ref. [21]).

Fig. 5.31. Resistance thermometers as part of Wheatstone bridge circuit: (*a*) uncompensated, (*b*) with compensation for variable lead-resistance.

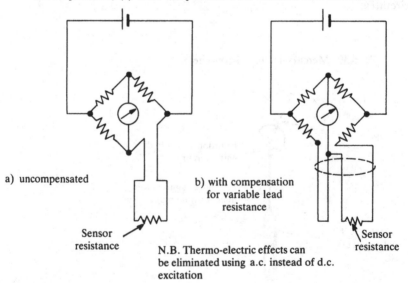

a) uncompensated

b) with compensation for variable lead resistance

Sensor resistance

Sensor resistance

N.B. Thermo-electric effects can be eliminated using a.c. instead of d.c. excitation

A *thermocouple* is a very simple temperature sensor which uses the Seebeck effect under zero current conditions to indicate the temperature of one junction relative to that of the other. Fig. 5.32 shows a simple thermocouple electric circuit in which the cold junction is maintained at a constant reference temperature of 0 °C by a bath of melting ice. The emf, V volts, is usually measured on a potentiometer. The thermocouple can be calibrated by measuring the emf at various known temperatures with the reference junction maintained at 0 °C. For most practical purposes the results of such measurements can, for most thermocouples, be represented with sufficient accuracy by the quadratic equation

$$V = A + BT + CT^2,$$

where A, B and C are constants relating to a particular thermocouple; and T is the temperature.

There is a large number of possible material combinations in thermo-electric thermometry but only a few of these are actually used in practice. These are chosen on the basis of thermoelectric potential, Seebeck coefficients, stability and reproductivity. Fig. 5.33 shows the Seebeck voltage plotted against temperature for a selection of typical material combinations. The end of each curve approximately indicates the upper limit of temperature for a particular combination. Some of these are marked with an asterisk to indicate they can be used also at low

Fig. 5.32. Thermocouple used as a thermometer.

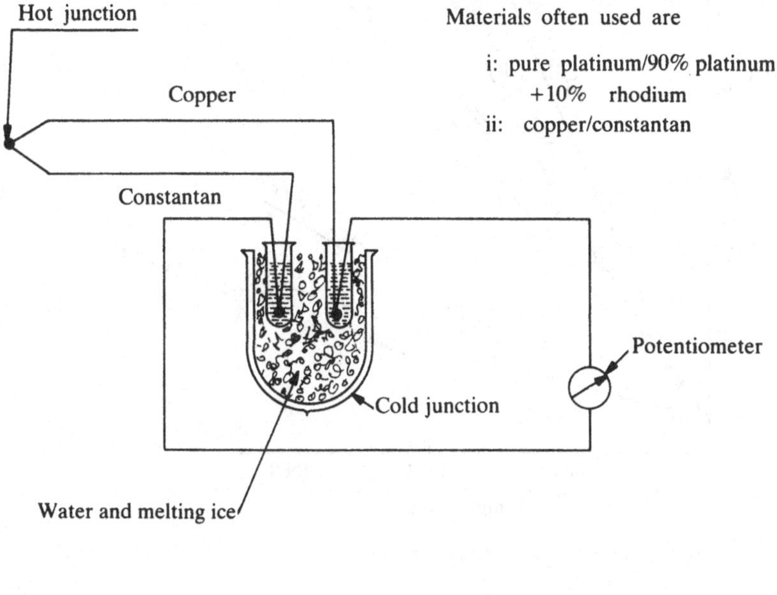

Hot junction

Copper

Constantan

Water and melting ice

Cold junction

Potentiometer

Materials often used are

i: pure platinum/90% platinum +10% rhodium

ii: copper/constantan

temperature ($-180\,^{\circ}$C). A useful discussion of thermocouples is given by Dally *et al.* (Ref. [9]).

Radiation detectors are used for measuring both radiant energy and temperature. They fall into two broad groups depending upon whether the radiation is measured over a wide band or a narrow band of wavelengths. In the case of the *total-radiation thermometer*, radiation emitted from a surface is focused onto a temperature-sensitive detector such as a thermocouple and the resulting electrical signal used to indicate temperatures up to about $3000\,^{\circ}$C.

Selective radiation thermometers, usually known as *optical pyrometers*, work as brightness comparators. One common type of optical pyrometer, see Fig. 5.34, consists essentially of a telescope containing a narrow band wave filter (usually red glass) and a tungsten filament electric lamp. When the current supplied to the lamp is gradually increased, the filament viewed against the brightness of the radiating body tends to disappear. A further

Fig. 5.33. Output voltage variation with temperature for various thermocouple combinations (reference junction temperature held at $0\,^{\circ}$C; * can be used down to $-180\,^{\circ}$C).

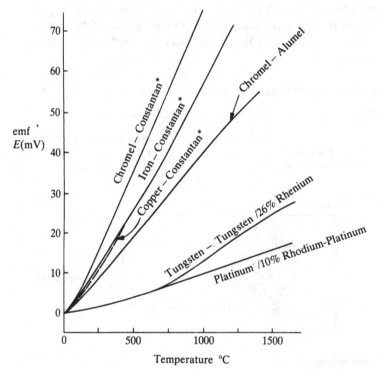

increase of current will cause the filament to glow more brightly than the background. If the relationship between heating current and filament temperature is known the temperature of the radiating surface can be determined. Commercial optical pyrometers often have the ammeter graduated to give a direct temperature reading.

The disappearing filament optical pyrometer can be used to measure the temperature of radiating bodies in excess of 700 °C up to the limit of the lamp filament at 1350 °C. When temperatures in excess of 1350 °C are to be observed, glass filters are interposed between the lamp and the radiating source to absorb some of the radiation. Temperatures in excess of 4000 °C can be measured in this way.

Optical pyrometers are calibrated by comparing the brightness of the tungsten filament with a blackbody source of known temperature or, more conveniently perhaps, by means of tungsten-strip lamps calibrated at NPL

Fig. 5.34. Optical pyrometer-disappearing filament type.

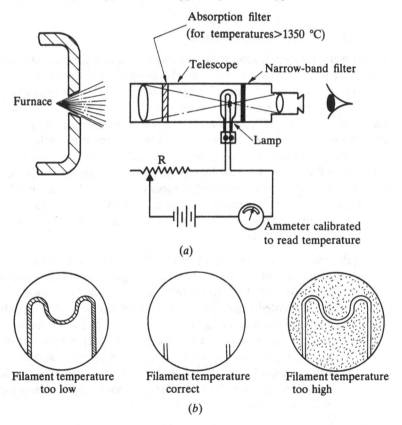

(a)

Filament temperature too low   Filament temperature correct   Filament temperature too high

(b)

or another similar laboratory. Blackbody behaviour (with emissivity $\varepsilon = 1$) can be closely simulated by a blackened conical cavity of about 15° included angle. Temperature is adjustable, controlled automatically and measured by some accurate sensor (e.g. a platinum resistance thermometer). According to Doebelin (Ref. [10]) a typical unit covers the range 500–1000 K with 1 K accuracy and $\varepsilon = 0.99 \pm 0.01$. The disappearing-filament pyrometer is the most accurate of all the radiation-type temperature-measuring devices and consequently it is used to establish the International Practical Temperature Scale above 1063 °C.

5.5.10 SUMMARY OF COMMONLY USED LABORATORY EQUIPMENT

In Table 5.2, the various items of equipment mentioned in this chapter have been tabulated, together with their range and resolution. For further details on any one item reference should be made to the foregoing text or to the list of references.

5.5.11 FLOW VISUALISATION

When an engineering structure fails, the location of that failure, and also the type of it, can usually be seen. This can also be so for even the most complex of structures and most catastrophic of failures. Thus observation can open the way to future prevention of this fault. In contrast, when there is a bad fluid flow it is often difficult to ascertain both the cause and its location. Thus resort is often made to means of visualising the flow pattern.

It is convenient in considering flow visualisation methods to distinguish between the flow adjacent to a solid surface and that forming a stream. Only a very few of the great number of techniques available can be described here. Extended texts are available for more details (Refs. [4], [15], [16] and [19]).

There are four main characteristics of surface flow which can be visualised as now described.

(a) The flow direction in air can be shown by fixing to the surface small tufts of thread. These will be aligned in the local direction of the flow by the aerodynamic force upon them. In water flows small pivoted vanes will show the flow direction. The introduction into a gas flow of a fine filament of smoke will trace out the flow direction: in water a line of coloured water will give the same result. A gas flow exerts a shear force on a surface and this will cause a liquid film, laid over the surface, to move in the flow direction. In water flow a suitably viscous liquid such as paint will also suffice.

Table 5.2. *Summary of commonly used laboratory equipment*

Parameter to be measured	Measuring device	Typical range	Approximate resolution	Remarks
Length	Machinist's calipers with steel rule	0–50 mm up to 0–1.0 m	±0.2 mm	Used mainly for checking nominal dimensions
	Vernier calipers or height gauge	0–0.1 m up to 0–0.5 m	±0.02 mm	
	Micrometer	0–25 mm or 25–50 mm, etc.	±0.005 mm	Range can be extended considerably to, say, 1 m ±12.5 mm by use of a standard 25 mm micrometer head and a 'bow gauge' structure
	Slip gauges	e.g. 0.5–100.0 mm	±0.001 mm	Used mainly for the calibration of lower precision instruments
Displacement	Dial gauge	e.g. 0–5 mm 0–50 mm	±0.01 mm	Extensively used, particularly in workshop practice
	Linear variable differential transformer	e.g. 0–0.25 mm up to 0–0.3 m	±0.1%	Resolution theoretically infinite but in practice dependent on associated instrumentation
Acceleration	Accelerometer	e.g 0–7000 g	0.05 g	With suitable instrumentation can be used for recording periodic displacements
Strain	Mechanical extensometers	e.g. Huggenberger type 0–0.4%	$\pm 10 \times 10^{-6}$	Range can be increased by resetting
	Electrical resistance strain gauges	±1.5% up to ±20% strain	$\pm 1 \times 10^{-6}$	Generally useful in transducer applications for measurement of displacement, load, torque, pressure, etc. Not re-usable and therefore costly

Table 5.2 (cont.)

Parameter to be measured	Measuring device	Typical range	Approximate resolution	Remarks
Time and frequency	Stopclocks	e.g. 0–60 min	±0.2 s	Not suitable for short periods of time, say, <10 s. Stopwatches available with better resolution
	Tachometer (mechanical or electrical)	e.g. 0–50000 rpm	–	Especially useful for shaft speed or frequency measurements, where it is un-desirable to have direct contact with the moving part
	Stroboscope	e.g. 0–100 Hz 0–300 Hz	–	By far the most widely used item of equipment for examination of transient and repetitive signals
	Cathode ray oscilloscope (cro) with transducer	e.g. 0.5 μs–1 s dc to 3 MHz	±5% –	Very versatile instrument for use with any form of wave or pulse pickup. Can be used for time or frequency measurement
	Electronic frequency counters with transducer	e.g. 1 μs–10^4 s 10 Hz–1.2 MHz	e.g. ±1 μs –	
	Trace recorders with transducer	0–2 Hz up to 0–5 kHz	±1%	Pen recorders are used for low frequency response; ultraviolet recorders, with suitable galvanometers, for high frequency response. Adjustable chart speed
Mass	Balances	e.g. 0–20 g → 0–200 g → 0–5 kg →	±0.001 mg ±1 mg ±4 mg	For extreme accuracy 'ultra-micro-balances' are available with a resolution of ±0.1 μg
Force	Spring balance	e.g. 0–20 N up to 0–5 kN	±1.0%	Resolution often poorer in the simpler spring balance arrangements
	Proving rings	e.g. 0–0.2 kN up to 0–70 kN	±0.1%	Accuracy dependent on technique used for measuring ring deflection

	Range	Accuracy	Comments
Strain gauge load cell	e.g. 0–40 N up to 0–1 MN	±0.1%	If not available commercially, strain gauge load cells can usually be devised to provide the required force measurement (see Table 5.1)
Hydraulic or pneumatic ram	Any range up to about 2 MN	See comments	Accuracy very much dependent on friction effects, but also on pressure-measuring system
Piezoelectric force transducer	e.g. 0–±8 kN up to 0–±5 MN	±20 mN	These transducers noted for their high resolution
Torque			
Prony brakes	General principle suitable for most laboratory power units	See comments	Used for power measurement in rotating shafts. Accuracy dependent on force measuring technique, etc.
Torsion dynamometer	0–10 mNm up to 0–5 kNm	±0.1%	This type suitable for either 'static' or 'rotating' torque measurement
Pressure			
U-tube manometer	20 mm–5 m of manometer fluid (vertical U-tube)	±0.5 mm ±0.05 mm at 5° to horizontal	Range can be increased but it is not usually practicable or convenient. Lower limit of range can be reduced to about 2 mm by tilting tubes to 5° from horizontal
Micromanometer	0–0.2 m of manometer fluid	±0.002 mm	A high-precision instrument
Bourdon tube	0–0.1 MN m^{-2} up to 0–500 MN m^{-2}	±1%	Combined negative and positive ranges sometimes used on low pressure gauges. Precision depends upon quality of device.
Dead-weight tester	100 N m^{-2} 100 MN m^{-2}	±0.01 to ±0.05%	Used for calibrating other gauges
Pressure transducers	0.002 mm of water to 500 MN m^{-2}	±0.1%	Sensitivity to low pressure differences limited by friction or by thermal expansion
Flow velocity			
Pitot-static tube	6–60 m s^{-1} usually	±1% of dynamic pressure	Quoted accuracy applies to International Standard designs. Possible to extend down to 1 or 2 m s^{-1}. Can be used up to Mach 5

Table 5.2 (cont.)

Parameter to be measured	Measuring device	Typical range	Approximate resolution	Remarks
	Hot-wire anemometer	Up to 250 m s^{-1} in filtered air at temperatures up to 150 °C. Up to 5 m s^{-1} in liquids	1 mm	Less accurate than pitot-static method except at very low velocities. Has very fast response – is very good for unsteady flow and turbulence measurements. Fragile
	Hot-film anemometer	50 m s^{-1} in gases 25 m s^{-1} in liquids	1 mm	Compared with hot wires they are much more robust, less susceptible to contamination but have a poorer frequency response
	Laser anemometer	—	0.2 mm cube	No physical interference in fluid but requires 'particles' 2–5 μm to give good Doppler signals
Flow direction	Wedge yawmeter	—	±0.2° or less	Also measures velocity. Works well in transverse gradients of total pressure. Small size
Mass or volume flow rate	Orifice plate	See remarks	±1%	Simple, cheap but has large losses in pressure. Size of plate or tube chosen to give a differential pressure of 0.5–5 m of fluid on a U-tube manometer
	Venturi meter	See remarks	±1%	Fairly expensive but has low pressure losses
	Vortex shedding flow-meter	1–30 m s^{-1} in liquids	±1%	Good repeatability, low cost, linearity, low pressure drop, rapid response

Temperature	Mercury in glass thermometer	−35–510 °C	±½°C	High temperature can cause ageing in glass. Zero calibration needed (in melting ice)
	Alcohol in glass thermometer	−80–100 °C		
	Pentane in glass thermometer	−200–30 °C		
	Mercury in steel thermometer with Bourdon gauge	−35–650 °C	±1%	Useful for remote readings. Correction for differences in level required
	Bimetal thermometer	−180–450 °C	±1%	Slow response
	Platinum resistance thermometer	−240–1060 °C	±0.2 °C (0 ≤ T ≤ 100 °C), ±0.35 °C (at 200 °C), ±0.55 °C (at 300 °C), ±0.8 °C (at 400 °C)	Very high precision – often used for interpolating between primary standard temperature points. See BS 1904 for temperature-resistance relationship and tolerances
	Thermocouples			
	Chromel and constantan	−180–1000 °C	±5 °C	Simple and cheap. Use in non-oxidising atmosphere
	Copper and constantan	−180–400 °C		
	Iron and constantan	−180–850 °C	±10 °C	
	Chromel and alumel	0–1100 °C	±10 °C	
	Tungsten and rhenium	0–2800 °C		
	Platinum and rhodium	0–1450 °C	±3 °C	Not subject to corrosion – very reliable – expensive. When calibrated against a standard.
	Optical pyrometers (disappearing filament type)	750–4000+ °C	±4 °C (at 1000 °C), ±6 °C (at 2000 °C), ±40 °C (at 4000 °C)	Glass absorption filter used above about 1350 °C
Voltage and current	Moving coil and moving iron instruments, etc.	See comments	±0.5% for precision grade ±1–±3% for industrial grade	Lower limit of effective range varies according to type of movement (typically this is 10% to 30% of full scale). See BS 89 for further information

General note: Any one instrument may cover only part of the ranges shown in the table. Ranges and resolution quoted are for guidance and may differ between manufacturers of the same type of instrument.

Fig. 5.35. The picture shows visualisation by smoke filament of the flow past a model car: the flow is from left to right. Separation is seen about half-way along the bonnet (hood), reattachment occurs half-way up the windscreen, and separation again follows at the front end of the roof. In the separation at the base of the windscreen there are two vortices which run across the bonnet (hood) and turn and continue along the side of the car. The flow behind the car is unsteady, as shown by the dispersion of the sharp smoke filaments.

(a)

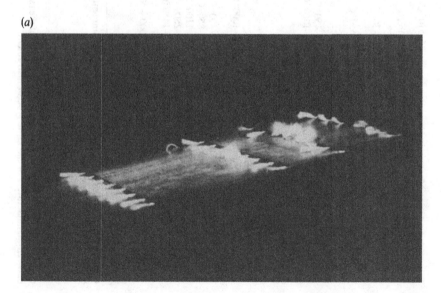

Fig. 5.36. This shows flow past and downstream of a plate inclined to the flow: (*a*) On the aerofoil the second row of tufts shows an unsteady flow downstream of separation with reattachment of the flow before the third row and steady flow again by the fourth row. At the far end of the second and third rows the rotation of these end tufts shows the presence of a vortex flow. (*b*) The plan view shows the tufts in the centre of the second row indicating the flow reversal downstream of separation. (*c*) The line of tufts across the flow downstream of the plate shows the vortex core rotation by the centre tuft and the external helical flow by the tufts on either side.

(*b*)

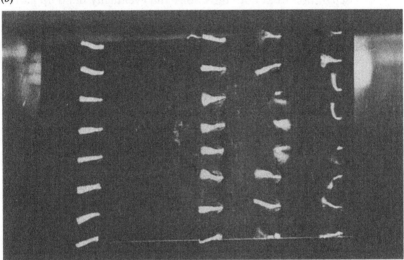

(*c*)

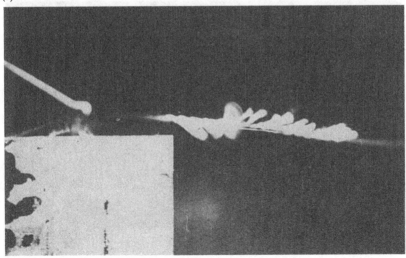

(b) The unsteadiness of a flow will be manifested as a fluctuation in the flow direction. Therefore the methods just described will also serve to distinguish between steady and unsteady flows.

(c) When separation of a flow occurs from a surface the flow adjacent to the surface just downstream of separation is reversed in direction. Thus again the previously described methods for showing flow direction can be used to visualise separation. In particular, a surface oil film will move towards the line of separation from both directions and eventually build up along that line as a fine ridge of oil.

(d) The breakdown of a steady, laminar flow into an unsteady, turbulent flow in the boundary layer adjacent to the surface is called transition. With care a smoke or dye filament will distinguish between the steady and the unsteady regions respectively up and downstream of transition. A film of oil will respond to the higher surface shear downstream of transition and the line of transition will eventually show as the division between the two films of markedly different thickness. If a suitable chemical is sprayed upon the surface then the higher rate of evaporation downstream of transition will eventually cause that portion to clear whilst leaving a film of liquid upstream.

Fig. 5.37. This shows, by smoke filaments, the flow from left to right over a concave surface. To the left stream vortices are shown by an evenly spaced set of smoke filaments which towards the right break up as the flow enters the transition to turbulence (photo by S. Riley).

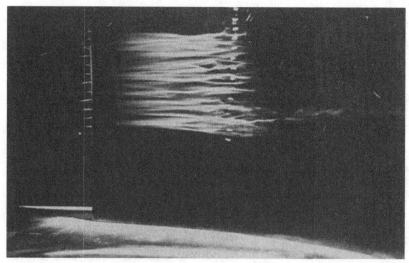

There are three principal characteristics of main stream flows that can be visualised and these are now described:

(a) The flow direction can be visualised as in the case of surface flows by using tufts, smoke and dye lines.

(b) Unsteadiness in a main stream can also be visualised in the same manner.

(c) A vortex flow is common; atmospheric whirlwinds and water-spouts are examples of these. The principle characteristic is the rotational nature of this flow. Both smoke and dye line will show up the rotating nature of a vortex by the appearance of a solid shaft of smoke or dye in the inner region and spiral filaments external to this. Tufts within the vortex core will have a rapid conical rotation whilst those in the external region will align along the helical path of the flow.

Illustrations of the techniques are given in Figs. 5.35–5.39, with detailed descriptions attached to these diagrams.

Just as with measuring instruments, the introduction of a visualising agent into a flow will also introduce some degree of disturbance to that flow. The skill in the use of visualisation is to cause no degree of disturbance that

Fig. 5.38. The photo shows an example of the 'china-clay' evaporation technique. The flow is from left to right over a circular cylinder. To the left the film of liquid has completely evaporated: to the right it remains so that its refractive index makes the light surface appear dark. This shows separation of the boundary layer under two conditions (photo by A. H. Yousif).

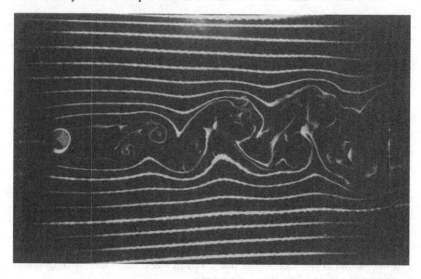

Fig. 5.39. This shows the flow from left to right past a circular cylinder set across the flow. A filament of smoke impinges upon the front of the cylinder and then passes each side showing separation about half-way round. It then picks up a regular series of vortex flows downstream. The flow away from the vortex flows is steady but becomes unsteady downstream in the vortex region.

will significantly falsify those features of the flow to be visualised. For example, tufts must be sufficiently small and smoke and dye lines must not be introduced as jets of too high a speed.

References

[1] Anonymous (1962). 'Balances, weights and precise laboratory weighing', *Nat. Phys. Lab. Notes on Applied Science*, No. 7.

[2] Anonymous (1979). *Methods and Practice for Stress and Strain Measurement*. Soc. for Strain Measurement.

[3] Anonymous (1978). *Tens-Lac: Brittle Coating for Stress-Analysis Testing*. Photoelastic Inc.

[4] *Symposium on Flow Visualisation*. ASME Annual Meeting (New York), 30 Nov. 1960.

[5] Benedict, R. P. (1969). *Fundamentals of Temperature, Pressure and Flow Measurement*. Wiley.

[6] Brooker, K. (ed.) (1984). *Manual of British Standards in Engineering Metrology*. BSI/Hutchinson.

[7] Bryer, D. W. & Pankhurst, R. C. (1971). *Pressure-Probe Methods for Determining Wind Speed and Flow Direction*. HMSO, London.

[8] Dally, J. W. & Riley, W. F. (1965). *Experimental Stress Analysis*. McGraw-Hill.

[9]   Dally, J. W., Riley, W. F. & McConnel, K. G. (1984). *Instrumentation for Engineering Measurements*. Wiley.
[10] Doebelin, E. O. (1983). *Measurement Systems, Applications and Design*. McGraw-Hill.
[11] Drury, J. D. (1978). *Ultrasonic Flow Detection for Technicians*. Quadrant Press.
[12] Evans, J. C. & Taylorson, C. O. (1964). 'Measurement of angle in engineering', *Nat. Phys. Lab. Notes on Applied Science*, No. 26, HMSO.
[13] Gibbings, J. C. (1970). *Thermomechanics*. Pergamon Press.
[14] Goodyer, M. J. (1974). 'A stagnation pressure probe for supersonic and subsonic flows', *Aero. Quart.*, **25**, 91–100.
[15] Holder, D. W., North, R. J. & Wood, G. P. (1956). 'Optical methods for examining the flow in high-speed wind tunnels', *Adv. Group Aero. Res. & Dev.*, Agardograph 23, Paris.
[16] Merzkirch, W. (1974). *Flow Visualisation*. Academic Press, New York.
[17] Mountain, D. S. & Webber, J. M. B. (1975). 'Stress pattern analysis by thermal emission (SPATE)', *Proc. 4th European Electro-Optics Conf.*, Vol. 164.
[18] Ower, E. & Pankhurst, R. C. (1966). *Measurement of Air Flow*. Pergamon Press.
[19] Pankhurst, R. C. & Holder, D. W. (1952). *Wind Tunnel Technique*. Pitman, London.
[20] Reynolds, O. (1882). 'On the internal cohesion of liquids ...', *Mem. Proc. Manchester Lit. Phil. Soc.*, 3rd Series, Vol. 7, pp. 1–19.
[21] Rogers, G. F. C. & Mayhew, Y. R. (1980). *Engineering Thermodynamics, Work and Heat Transfer*. Longman.
[22] Ryley, D. J. (1980). 'Hydrostatic tensile stress in water', *I.J.M.E.E.*, **8**(2).
[23] Window, A. L. (1982). *An Introduction to Strain Gauges*. Welwyn Strain Measurement Ltd.

6
Photography in experiments

6.1 Introduction (a list of the principal notation used in this chapter is given on pp. 198–199)

The use of photography in experiments is not only common but is widespread in the nature of its application. In general terms it is used to illustrate the experiment and to record events. For both of these it is important to realise that the criteria for a competent application can differ markedly from those for either artistic or journalistic photography. The use of photography in experiments is a means of communicating scientific information. All the information to be transmitted must be shown with equal clarity: in contrast, artistic photography commonly and deliberately conceals some information in shadows and emphasises others in highlights. This is a difference that is important.

Despite a belief to the contrary held by many students, there is often no justification for preparing a photograph of experimental apparatus. Commonly a superior form of presenting the information is by a line diagram that has been carefully composed and drawn. When a photograph is prepared some care should be taken in composing the object. The view should not be confused by either large and irrelevant pieces of equipment in the foreground or by extraneous wiring, tubing and glassware, all of which mask or distract from the information presented.

When photography is used as a means of measurement then the requirement of accuracy is over-riding. For example, a picture from a cathode ray oscilloscope would need to show a thin trace for precision of measurement. When, in contrast, such a picture is used to illustrate a phenomenon in qualitative terms then a thicker trace may be preferable for clarity. This is illustrated by the two photographs of Fig. 6.1. The first picture (*a*) shows a thin trace to enable the frequency of the fairly smooth waves to be measured; the second, (*b*), shows a thick trace which reveals all the trace of the burst of random behaviour which was partly missing in the first.

Fig. 6.1. Two photographs to illustrate the different requirements of illustration and measurement: (*a*) fine trace for measurement of wave length which misses some detail; (*b*) thick trace for illustration.

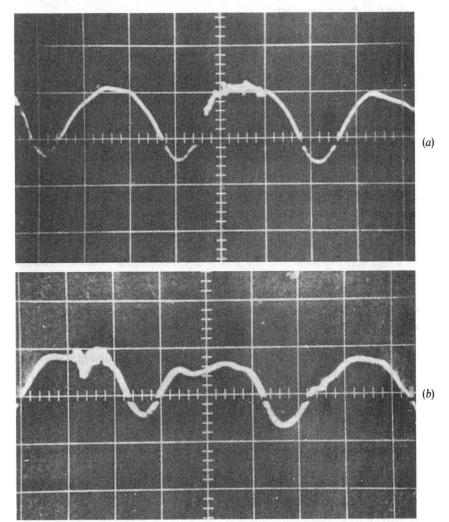

(*a*)

(*b*)

Table 6.1. *Paper sizes*

Print paper (in)	Print paper (mm)	Standard size		Page size	
		Number	Size (mm)	Number	Size (mm)
$3\frac{1}{2} \times 5$	89×127	B7	88×125	A5	149×210
5×7	127×178	B6	125×177	A4	210×297
8×10	203×254	B5	177×250	A3	297×420

Table 6.2. *Paper diagonals*

Standard number	Diagonal size (mm)	Standard number	Diagonal size (mm)
B7	153	A5	257
B6	217	A4	364
B5	306	A3	515

6.2 Picture composition

The great majority of reports and theses are produced on a standardised sheet size known as A4 size. The two reference sizes are A0 having an area of 1 m^2 and B0 of area $\sqrt{2} \text{ m}^2$. Both standards have sides in the ratio of $\sqrt{2}:1$ so that when successively folded the resulting sheets retain this proportion. The three most frequently used sizes of photographic print paper are listed in Table 6.1, together with the metric equivalents, the nearest standard equivalent and the suggested equivalent page size of a report. This table implies a generous allowance for a border around a print: as well as being advisable for addition of figure captions and to allow binding of a manuscript, the proportions of pictures are then near to what the ancients laid down as an artistically ideal proportion known as the 'golden-mean'; that is, the proportion $(1 + \sqrt{5})/2 = 1.62$. However, the demands of scientific communication often prohibit adherence to these ideal shapes of print and it is common to print on oversized paper and trim to a required size.

The largest angle over which the eye can comfortably move is of a solid angle, ω, of about $\frac{2}{3}$ sr. The equivalent cone angle, 2θ, is given by

$$\omega = 2\pi[1 - \cos \theta], \tag{6.1}$$

and so is $53°$. A typical reading distance is 40 cm and so a diagonal size based on this cone angle is $2 \times 40 \times \tan (53/2) = 40$ cm. The diagonal size for the various standard paper sizes are given in Table 6.2. Use of A4 paper size is seen to lie comfortably within the limit to the diagonal size.

Table 6.3. *Frame size (mm)*

Film type	Height	Width	Diagonal (w_f)	Width / Height
8 mm cine	3.75	5.0	6.25	1.33
16 mm cine	7.5	10.0	12.5	1.33
35 mm still	24.0	36.0	43.3	1.50
120 square still	60.0	60.0	84.9	1.00
120 still	60.0	82.0	101.6	1.37
'Polaroid' still ($\frac{1}{4}$ plate)	83.0	108.0	136.0	1.30
Technical camera	100.0	125.0	163.0	1.25

The picture is formed on film and the frame sizes for the most commonly used film are given in Table 6.3.

6.3 Picture precision

(a) FOCUS AND MAGNIFICATION

The standard lens formulae are

$$\frac{1}{u}+\frac{1}{v}=\frac{1}{f} \tag{6.2}$$

and

$$m=v/u, \tag{6.3}$$

where m is the magnification and the variables are illustrated in Fig. 6.2.

(b) VIEWING ANGLE

A requirement for a camera lens is that the viewing angle should be the same as the viewing angle of the resulting print. This gives the correct perspective to the picture. However, this requirement can be relaxed to a degree without the distortion of perspective being readily apparent. Inspection of Fig. 6.2 shows that the viewing angle, θ, is given by

$$\tan\theta=\frac{w_f}{2v},$$

where w_f is the film diagonal size. Using equations (6.2) and (6.3) gives

$$\tan\theta=\frac{w_f}{2f(1+m)}, \tag{6.4}$$

and when, as is often the case, $m \ll 1$, then

$$\tan \theta = w_f/2f. \tag{6.5}$$

As an illustration, two numerical examples are given in Table 6.4. In these, the paper sizes chosen from Table 6.1 and detailed in Table 6.2 are illustrative, as are the film sizes detailed in Table 6.3. With the object sizes as chosen then the lens requirements from equation (6.4) are obtained in the final row of this table.

(c) LENS APERTURE EFFECTS

The effective diameter of the lens is adjusted by a lens stop which thus controls the amount of light falling on the film. The lens stop number, A, is usually quoted as f/A, e.g. $f/8$. The number is the ratio of this effective diameter, d_1, to the focal length, f, or

$$A \equiv d_1/f. \tag{6.6}$$

As the light transmitted by the lens will be proportional to the area of the opening, or to A^2, then stop numbers follow the sequence listed in Table 6.5. The sequence is seen to follow one of integral values of $\log_2 A^2$. The numerical scale set on a lens is thus a logarithmic one and uses the values

Table 6.4. *Numerical examples*

Case	(a)	(b)	Source
Paper size	B7	B5	Table 6.1
Paper diagonal, w_p mm	153.0	306.0	Table 6.2
Viewing angle	21.7°	41.9°	Distance 40 cm
Film size	35 mm	120	
Film diagonal, w_f mm	43.3	101.6	Table 6.3
Object diagonal, w_o mm	1000.0	5000.0	
Magnification, m	4.33×10^{-2}	2.03×10^{-2}	w_f/w_o
Lens focal length, f mm	109.0	130.0	Equation (6.4)

Fig. 6.2. Optical parameters.

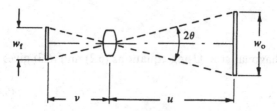

Table 6.5. *Lens stop numbers*

Sequence number	A	Approximate A	$\log_2 A^2$
0	1	1	0
1	$\sqrt{2}$	1.4	1
2	2	2	2
3	$2\sqrt{2}$	2.8	3
4	4	4	4
5	$4\sqrt{2}$	5.7*	5
6	8	8	6
7	$8\sqrt{2}$	11	7
8	16	16	8
9	$16\sqrt{2}$	23*	9

* Note: For some inexplicable reason the values of A in the sequence of 5 and 9 have always been quoted incorrectly as 5.6 and 22.

tabulated as 'approximate A'. The difference between two successive sequences is referred to as 'one-stop'. Then a proportion of a stop, r, added on to one of A_0 gives a proportional increase in light, p, of

$$r = \log_2(1+p). \tag{6.7}$$

For example, an increase of $\frac{1}{3}$ of a stop in the sequence gives an increase of light by the fraction 0.26 and the value of A is increased in the ratio $1.26^{\frac{1}{2}} = 1.12$. The resolution of detail of the image is measured as 'lines per millimetre', n, or the line spacing, s, where $s = 1/n$. Rayleigh's formula for this is (Ref. [1])

$$s = 1.22\lambda A,$$

where λ is the wavelength of the light transmitted. Taking $\lambda = 500$ nm then

$$n = 1.64 \times 10^3/A \quad \text{lines mm}^{-1}.$$

Real lenses do not achieve this precision of reproduction and typical measured values are shown in Fig. 6.3. It is seen that the resolution is better at the centre of the image than at the edge. Values peak at about $A = 8$. To this graph is added Rayleigh's value reduced by 21% and also a curve proposed by de Vaucouleurs as being typical of reasonably good-quality lenses (Ref. [2]).

There is also a limitation on the precision of resolution by film emulsion. Some typical values are shown in Fig. 6.4 for black-and-white film. The resolution is seen to depend upon the degree of contrast in the image and upon the film speed, S_A.† The resolution is improved by use of a film of

† A topic discussed in detail later.

reduced speed. It should be noted that Fig. 6.4 provides only a guide as manufacturers continually improve the resolution of their emulsions.

Consider firstly the combined effects of resolution in the film emulsion and by the camera lens. If the resolved number of lines per millimetre for the film and lens respectively are n_f and n_l then the respective spacings are $1/n_f$ and $1/n_l$. The spacing $1/n_l$ is represented by the horizontal vector shown in Fig. 6.5 along the line Oa. The spacings $1/n_f$ and $1/n_l$ are akin to random errors. Then it is supposed that there is a zero probability of $1/n_f$ being either exactly in line or exactly opposite to $1/n_l$: these are the directions Oa and Oc in Fig. 6.5. Further, it is supposed that the maximum probability is for $1/n_f$ to lie half-way between these two positions: that is, the direction Ob in Fig. 6.5. Then the combined error in spacing is taken as the vectorial sum along Od. This is $1/n_1$, given by

$$\frac{1}{n_1} = \left[\frac{1}{n_l^2} + \frac{1}{n_f^2} \right]^{\frac{1}{2}}. \tag{6.8}$$

Fig. 6.3. Lens resolution.

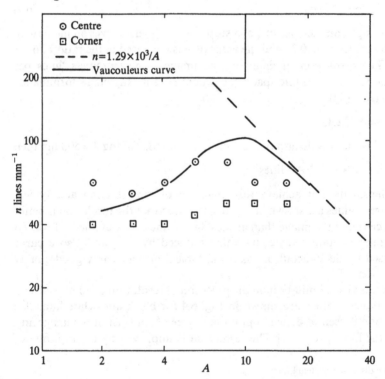

On enlarging from a film of width w_f to a print of width w_p the above spacing error will become

$$\frac{w_p}{w_f}\frac{1}{n_1} = \frac{w_p}{w_f}\left[\frac{1}{n_l^2} + \frac{1}{n_n^2}\right]^{\frac{1}{2}}.$$

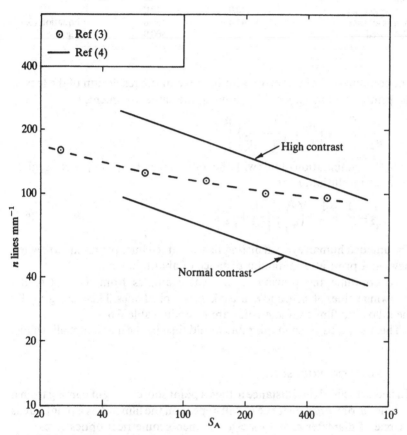

Fig. 6.4. Film resolution.

Fig. 6.5. Combination of errors of resolution.

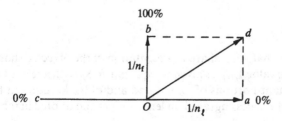

Table 6.6. *Film speed examples*

Case	(a)	(b)	Source
Lens resolution, n_1	80.0	60.0	Fig. 6.3
Enlarger resolution, n_e	80.0	80.0	
Print resolution, n_p	40.0	40.0	
Film resolution, n_f	41.4	36.8	Equation (6.9)
Film speed, S_A	570.0	800.0	Fig. 6.4

The combination of this error with that due to the resolution of the lens of the print enlarger, n_e, would, following the same argument, be,

$$\frac{1}{n_2} = \left[\frac{1}{n_e^2} + \left(\frac{w_p}{w_f}\right)^2 \left\{\frac{1}{n_1^2} + \frac{1}{n_f^2}\right\}\right]^{\frac{1}{2}}.$$

Then the combination of this with the resolution of the print paper, n_p, gives the final resolution, n, as

$$\frac{1}{n^2} = \frac{1}{n_p^2} + \frac{1}{n_e^2} + \left(\frac{w_p}{w_f}\right)^2 \left[\frac{1}{n_1^2} + \frac{1}{n_f^2}\right]. \tag{6.9}$$

The unaided human eye can distinguish about ten lines per millimetre when viewing a print at a minimum distance of about 100 mm.

To continue the previous numerical examples from Table 6.4, the following values of n_1, n_e and n_p are taken as typical ones. Then, using $n = 10$, the following film characteristics are given in Table 6.6.

The film speed given in this table would then be the maximum allowable.

(d) LENS FOCUSING

If a lens is focused on a distance u, then a point source of light coming from a distance u' or from a distance u'' will appear on the film not as a point but as a circle of diameter c. With $u' > u > u''$ then geometrical optics gives,

$$A = \frac{f^2}{c} \frac{(u' - u'')}{2u'u'' - f(u' + u'')} \tag{6.10}$$

and

$$u = \frac{2u'u''}{u' + u''}. \tag{6.11}$$

It is clearly desirable that, for a sharp representation of the object, c should not be more than the value $1/n_1$, as given by equation (6.8), with a relaxation of this for unimportant portions of foreground and of background in the picture. Depth of field scales engraved on lens mounts and quoted in tables

Table 6.7. *Stop number examples*

Case	(a)	(b)	Source
u' mm	2750.0	7000.0	
u'' mm	2000.0	6000.0	
n_1 lines mm^{-1}	36.8	31.40	Equation (6.8)
c mm	5.44×10^{-2}	6.38×10^{-2}	Equation (6.12)
u mm	2316.0	6462.0	Equation (6.11)
v mm	108.0	130.0	Equation (6.2)
A	20.1	3.4	Equation (6.10)
A_0	16.0	$2\sqrt{2}$	
r	0.66	0.53	Equation (6.7)

have been determined for a fixed value of c. It has been a common practice (Ref. [3]) to take $c = f/1000$ for still cameras and $c = f/1500$ for cine cameras. A more acceptable practice is to use the first-mentioned criterion but to relax it by taking (Ref. [3])

$$c = 2/n_1. \tag{6.12}$$

Continuing the previous numerical example in Table 6.6 gives the following results of Table 6.7 for the values of u' and u'' specified. The last two entries in Table 6.7 show how the lens stop is to be set on its logarithmic scale. For case (a) it would be set at 0.66 of a stop up from $f/16$.

In setting the lens to focus at the distance u it can often help precision if an object, such as a fine wire, is placed in a suitable location by the object to be finally photographed. Where an image is to be focused on a ground-glass screen, the image of such a wire should be viewed through a magnifying lens. An illustration of obtaining adequate depth of field is given in Fig. 6.6(a) and (b). Full details are given in the title to this diagram. Fig. 6.6(a) shows adequate depth of field; (b) shows an inadequate one. In certain cases where it is not possible to obtain an adequate depth of field in accordance with equation (6.10), then a camera with a tilting back can be used, as illustrated in Fig. 6.7(a). Provided the axis of the cone of the view is perpendicular to the lens plane then an exact focus over the whole negative can be obtained. It is readily shown geometrically that the requirement for this is that the planes of the object, lens and film should all intersect along a common line as shown in this diagram. The drawback is that the perspective is distorted on the film.

When a camera cannot be positioned opposite an object then a distorted perspective results. This can be overcome by using a camera with a rising front, as illustrated in Fig. 6.7(b). Set up as shown, all the vertical parallel lines will now record also as parallel on the film.

Fig. 6.6. These photographs are of four sets of equipment for undergraduate experiments on jet flow. The nearest unit was at $u'' = 2$ m, the farthest at $u' = 10$ m. The lens used was of $f = 50$ mm. Taking $c = 2/40$ mm, equation (6.10) gives $A = 10.1$ and equation (6.11) gives $u = 3.3$ m. Fig. 6.6(a) was taken at $A = 16$ and $u = 3.3$ m; Fig. 6.6(b) was taken at $A = 1.7$ and $u = 3.3$ m. The luminance of a vertical white card at the nearest and farthest unit was 30.9 and 43.7 cd m^{-2}. The corresponding readings from a 'Weston' meter using an 'invercone' were 4.8 and 5.9 cd m^{-2}. $S_A = 400$ and for (a) $t = 1$ s; (b) $t = \frac{1}{60}$ s.

(a)

(b)

6.4   Exposure time

There are two forms of camera shutter, the 'focal-plane' type and the 'between-lens' one.

The 'focal-plane' shutter consists of a roller blind with a gap in it whose width can be adjusted. The blind is situated just in front of the film and runs across it at a constant velocity so that the gap passing across exposes the film progressively. The exposure time is effectively varied by varying the gap width. The system is sketched in Fig. 6.8(*a*). This type of shutter can be unsatisfactory for scientific work when an object is in motion, for then the shape of the image will be distorted if the total time for the gap to expose the whole width of the film is comparable with the time of significant movement of the image at the film plane.

Fig. 6.7. Special use of adjustable technical cameras.

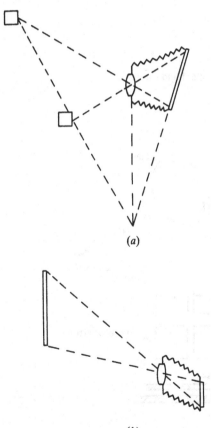

(*a*)

(*b*)

The 'between-lens' shutter is an opening that moves across the light path near to the lens stop. A simple model is shown sketched in Fig. 6.8(b). As the shutter opens there are three stages of illumination at the film plane. Firstly there is a gradual initial rise in illumination until the lens aperture is completely exposed; then if this aperture is smaller than the open area of the shutter there follows a period of constant illumination; finally there is a gradual fall in illumination as the lens aperture is gradually covered.

With the simplified shutter system of Fig. 6.8(b) the shutter of fixed opening width w, is moving at a constant, but variable velocity, V, across a square stop opening of side d that is variable. The illumination, E, at the film plane can be equated to kd^2. The variation of illumination with time follows the pattern, shown in Fig. 6.9(a), which is illustrated for two stop openings

Fig. 6.8. Sketches of camera shutters: (a) roller blind – focal plane; (b) between lens.

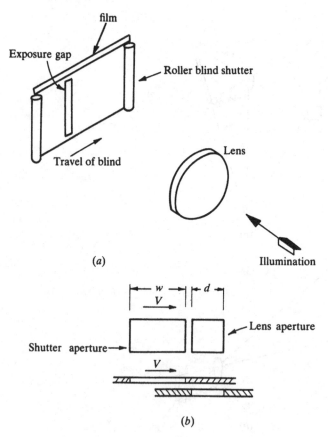

d_1 and d_2. The pattern for two shutter velocities, V_1 and V_2, is shown in Fig. 6.9(b). The exposure, H, is given by the integral during the opening time, t, as

$$H = \int E \cdot dt \qquad (6.13)$$

and the effective exposure time, t_f, is defined by

$$t_f = \frac{1}{E_m} \int E \, dt,$$

where E_m is the maximum value. Inspection of Fig. 6.9(a) and (b) shows that

$$t_f = w/V,$$

Fig. 6.9. Illumination during shutter opening: (a) constant shutter speed; (b) constant aperture setting.

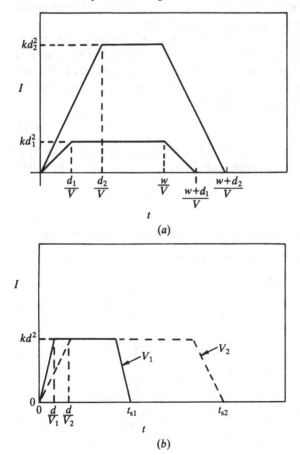

(a)

(b)

and so the effective exposure time is independent of aperture setting. The reference shutter time (speed), t_e, which forms the numerical scale attached to shutters, is commonly taken as the time between the half-rise and fall times at full aperture. Again inspection of Fig. 6.9(a) and (b) shows that $t_e = t_f$ for this simple form of shutter.

For other shutter shapes the effective shutter speed, t_f, can differ from t_e. With circular openings of both shutter and lens stop the rise and fall rates are no longer constant with time t even for a constant shutter speed. Calculation gives the result shown in Fig. 6.10: it was done for a stop number of unity when both apertures are of the same diameter, and for a constant shutter velocity. The variation of effective shutter speed with change in stop number is shown in Fig. 6.11. Also shown are some reported experimental values (Ref. [5]). Whilst there is good agreement between the experimental and calculated variation at the slower shutter speeds, for short times there is a large discrepancy; the actual shutter is slower at short speeds and small stop apertures by a factor of nearly two.

When the object is moving, the exposure time must be chosen to effectively 'freeze' that motion on the film. If the object is moving at a velocity V at an angle of α to the camera axis then, with reference to Fig. 6.2, the velocity of the image across the film will be $(v/u) \cdot V \sin \alpha$ so that the distance moved on the negative will be

$$(v/u) \cdot V \cdot t_e \cdot \sin \alpha. \tag{6.14}$$

Because the object, of size s_0, is moving towards the camera at a velocity of $V \cos \alpha$ then the value of u will change by $-Vt_e \cos \alpha$. The size of the image,

Fig. 6.10. Relative illumination during shutter time (A: stop number).

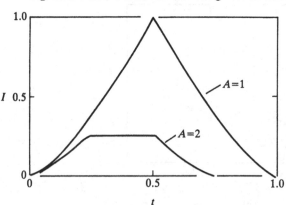

s_i, will correspondingly change. So that, using equation (6.3),

$$m \equiv \frac{s_i}{s_o} = \frac{v}{u},$$

then

$$\frac{ds_i}{s_o} = -\frac{v}{u^2}\, du = \frac{v}{u^2}\, Vt_e \cdot \cos \alpha.$$

Fig. 6.11. Effective shutter exposure time.

⊙ Experiment (Horder)

—— Calculation

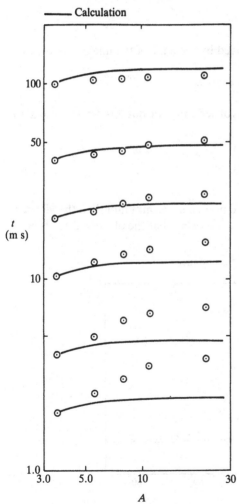

If the object subtends an angle 2θ at the lens then

$$\frac{s_o/2}{u} = \tan \theta. \tag{6.15}$$

The edge of the image will then move a distance

$$\frac{ds_i}{2} = \frac{1}{2}\frac{s_o}{u}\frac{v}{u}\, Vt_e \cdot \cos \alpha$$

$$= \frac{v}{u} \cdot Vt_e \tan \theta \cdot \cos \alpha.$$

The total maximum distance moved by the edge of the image will then be

$$(v/u)Vt_e \sin \alpha + (v/u)Vt_e \cos \alpha \tan \theta.$$

To accord with the previous limits of definition we put this distance equal to $1/n_1$ and so

$$t_e = \frac{u}{v} \cdot \frac{1}{n_1 V}\bigg/ [\sin \alpha + \cos \alpha \tan \theta]. \tag{6.16}$$

As an example, for a lens of $f = 50$ mm using 35 mm film then the viewing angle from equation (6.5) is $46.8°$. Supposing that the object subtends 60%

Fig. 6.12. Exposure time to freeze image.

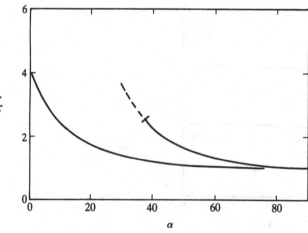

Table 6.8. *Shutter times examples*

Case	(a)	(b)	Source
2θ	13.0°	25.1°	Equation (6.15)
$\tan\theta$	0.114	0.223	
α	90°	30°	
V m s^{-1}	2.5	10.0	
t_e s	2.05×10^{-3}	2.28×10^{-4}	Equation (6.16)
$1/t_e$ s^{-1}	488.0	4378.0	

of this angle, then equation (6.16) gives the values shown plotted in Fig. 6.12. These results agree closely with published recommended rules (Ref. [3]). Following the preceding numerical examples of Tables 6.4, 6.6 and 6.7, retaining the 60% proportion mentioned above, and specifying values of α and V, gives the values in Table 6.8.

In case (a) a shutter speed of $\frac{1}{500}$ s would be suitable. In case (b), flash photography would have to be used to give a flash of duration of $\frac{1}{5000}$ s.

An illustration of image freezing is given in the photograph of Fig. 5.35. The vapour trails to illustrate this flow were moving across the line of view at about 5 m s^{-1}. The optical distances were $u = 3$ m, $v = 50$ mm. Taking $n_1 = 40$ mm^{-1}, then equation (6.16) gives $t_e = 3 \times 10^{-4}$ s. Two flash units were mounted in the side walls of the wind tunnel having flash times of about 5×10^{-4} s. The photograph shows that the motion was satisfactorily 'frozen'.

It is important that a camera be mounted on a suitably stiff tripod or other form of support. If there are over-riding reasons why a camera should be held by hand then 'natural' camera shake must be taken into account. The motion of the operator's hand can be taken typically as an angular velocity of about 1.5° s^{-1}. This corresponds to an object velocity, V, given by

$$\frac{V}{u} = \frac{1.5\pi}{180}.$$

Then from equation (6.14), taking $n_1 = 40$ mm^{-1} and $\alpha = 90°$ gives

$$vt_e = \frac{u}{Vn_1} = \frac{180}{1.5\pi \times 40}.$$

When $u \gg v$ then $v \simeq f$ and so

$$ft_e = 0.95. \tag{6.17}$$

This gives the following examples:

f mm	t_e sec eq. (6.17)	t_e sec shutter setting
50	1.9×10^{-2}	1/60
400	2.4×10^{-3}	1/500

This shows that the maximum acceptable exposure time can be an important restriction.

6.5 Lighting

Fig. 6.13 shows a point source of light emitting a luminous flux, F lm. The intensity of this light, I, is defined by

$$I \equiv df/d\omega \quad \text{lm sr}^{-1}, \tag{6.18}$$

where ω sr is the solid angle. This unit of lm sr^{-1} is the candela. The resulting illumination, E, on a surface of area δa is defined by

$$E \equiv dF/da \quad \text{lm m}^{-2}. \tag{6.19}$$

This unit of lm m^{-2} is the lux. When dA is an element of area perpendicular to the direction of the intensity, then as

$$\delta A = \delta a \cos \theta,$$

so also,

$$E = \frac{dF}{dA} \cos \theta \, \text{lx}. \tag{6.20}$$

Fig. 6.13. Illumination from a uniform point source.

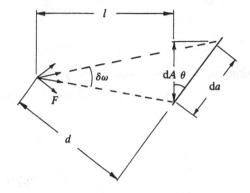

From Fig. 6.13 we have

$$da \cos \theta = l^2 \, d\omega,$$ (6.21)

so that using equations (6.18), (6.19) and (6.20) gives

$$E = I \cos \theta / l^2,$$

and from Fig. 6.13

$$E = I \cos^3 \theta / d^2.$$ (6.22)

A mean value of the illumination, $\bar{E}$, can be defined by

$$\bar{E}a = \int E \, da.$$ (6.23)

Fig. 6.14 shows the case of a surface of circular shape aligned so that the light impinges vertically at the centre. If the light source is both effectively centred as shown and is also uniform, then with

$$da = 2\pi r \, dr$$

and

$$r = d \tan \theta,$$

so that

$$dr = d \sec^2 \theta \, d\theta,$$

we find

$$a\bar{E} = \frac{I}{d^2} \int \cos^3 \theta \, 2\pi r \, dr$$

$$= \frac{I}{d^2} \int_0^{\theta_0} 2\pi d^2 \sin \theta \, d\theta$$

$$= 2\pi I (1 - \cos \theta_0).$$

Fig. 6.14. Illumination of a circular area.

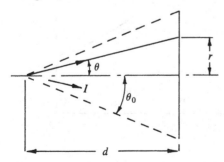

This can be rewritten,

$$\frac{\bar{E}d^2}{I} = \frac{2(1 - \cos\theta_0)}{\tan^2\theta_0},$$ (6.24)

whilst

$$\sqrt{a/d} = \sqrt{\pi}\tan\theta_0.$$ (6.25)

Values are shown plotted in Fig. 6.15. For a surface of rectangular shape, as shown in Fig. 6.16, then

$$a\bar{E} = 4Id \int_0^{w_1}\int_0^{w_2} \frac{dx\,dy}{(d^2 + x^2 + y^2)^{\frac{3}{2}}}.$$

This integrates to give

$$\frac{\bar{E}d^2}{I} = \frac{d^2}{w_1 w_2}\tan^{-1}\left[\frac{w_1 w_2/d^2}{(1 + (w_1^2 + w_2^2)/d^2)^{\frac{1}{2}}}\right],$$

Fig. 6.15. Distribution of illumination on surfaces.

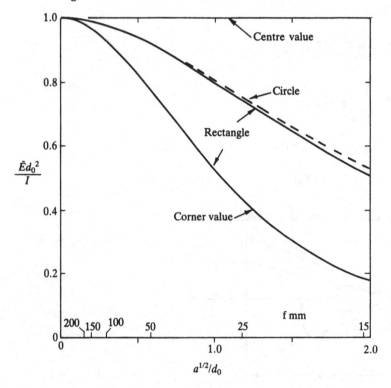

whilst

$$\sqrt{a/d} = 2(w_1 w_2)^{\frac{1}{2}}/d.$$

Values in Fig. 6.15 are for the case $w_1/w_2 = 1.5$ which is the ratio of the frame size for 35 mm film. Values for the square shape given by $w_1 = w_2$ would be barely distinguishable in this diagram. Also shown are the values of the illumination at the centre of each surface and also at the corners of the rectangular surface.

Inspection of this diagram shows that the mean illumination is closely similar for all three shapes. The distribution of illumination becomes seriously non-uniform with increase in the value of $\sqrt{a/d}$, that is, with closeness of the light sources to the surface. This latter quantity can be related to the focal length of a lens just viewing the surface and situated closely adjacent to the light source. For the surface viewed on 35 mm film of frame size given in Table 6.3, and using equation (6.5), a scale of focal length has been set out in Fig. 6.15. It is seen that using a 25 mm lens in these conditions means that a single light source gives an illumination at the corner that is only 43% of that at the centre. For purposes of scientific communication of information, a true photographic representation requires a uniform illumination and hence diffuse lighting rather than the point source just considered.

Uniform illumination can be approached by suitable reflecting and diffusing surfaces or by a number of light sources.

Equation (6.22) gives the distribution of light from a single source. It is a law of optics that values of the luminous flux from various sources can be added to give the total value. Then if the flux from one source on the area δa is δF_i then the total flux δF is given by

$$\delta F = \sum \delta F_i.$$

Fig. 6.16. Illuminated rectangular surface.

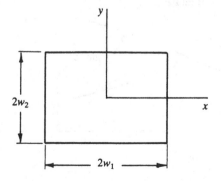

Noting equation (6.19) it follows that by dividing the above equation through by δa then

$$E = \sum E_i,$$

where E_i are the contributions from individual sources.

The combination of two light sources, each a distance d from a plane surface and spaced a distance $2w$ apart, is illustrated in Fig. 6.17. The two plots of equation (6.22) are shown in Fig. 6.18 together with the sum of these two. These graphs are drawn for the case of $2w/d = 1.20$. It is readily shown from application of equation (6.22) that this results in equal values of Ed^2/I at positions marked a, b and c in Fig. 6.17: this value is 1.26. The distribution is seen to be uniform to within $\pm 4\%$ over the range of $y/d = \pm 0.725$.

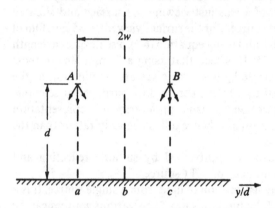

Fig. 6.17. Diagram of two light sources illuminating a plane surface.

Fig. 6.18. Distribution of illumination from two sources.

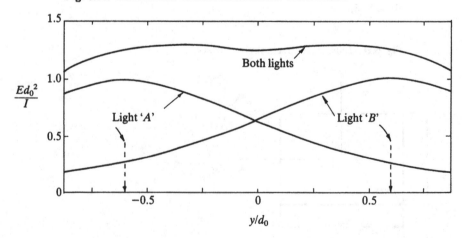

The distribution over the whole area of the surface illuminated can be obtained similarly by addition of the values of Ed^2/I. This is illustrated in Fig. 6.19. The left-hand side of this diagram shows two contours of constant Ed^2/I given by each light; one is for a value of 0.1 from that situated at position A, the other is for a value of 0.2 from that at B. The intersect gives the combined value of 0.3. In this way the contours shown in the right-hand half of this diagram can be constructed. Inspection of these contours shows that if it is judged acceptable that the illumination should lie somewhere in the range $1.0 < Ed^2/I < 1.3$ then a rectangular surface of about $(d) \times (1.5d)$ can be lit satisfactorily; if the size of the surface is specified then the value of d follows. It is convenient that this proportion of 1.5:1 is that of the frame of 35 mm film quoted in Table 6.3: the other frame proportions would be less acceptably illuminated.

It is seen from Fig. 6.17 that no specification of a direction of the lighting is made. For lights mounted in reflectors as used in photographic work data

Fig. 6.19. Distribution of illumination from two sources over a plane surface.

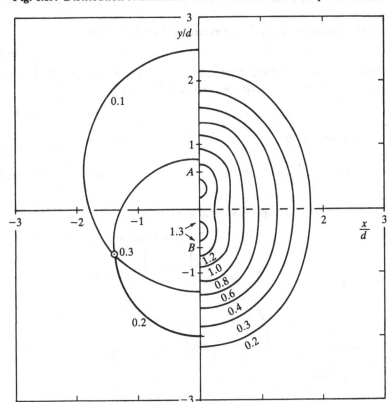

show that these lights obey the $1/d^2$ relation of equation (6.22) (Ref. [6]). Thus Fig. 6.19 is still valid. Then Fig. 6.17 indicates that the result of Fig. 6.19 applies whether the axes of the reflectors are directed towards point (b) or separately to (a) and (c) except that the beam will be of an angle ω sr with a cut-off of the light outside this angle. From the data referred to, typically $\omega = 0.27\pi$ sr. For this beam angle, two lights at $y/d = \pm 0.6$ cover a range of $y/d = \pm 0.7$ and $x/d = \pm 0.6$ when the axes of the beams are directed at the points $y/d = \pm 1/6$.

6.6   Reflectance and luminance

The vast majority of surfaces to be photographed reflect light over a spread of directions, unlike the perfect mirror which reflects all the light in a direction related uniquely to the incident direction. When a surface reflects light in all directions then the luminance of the surface is represented by the sketch of Fig. 6.20. This illustrates a small flux of light, δF, falling upon the area δa. A portion $\delta(\delta F)$ is then reflected at the angle θ shown: as δa is to be infinitesimal then the quantity $\delta(\delta F)$ is effectively issuing from a point source.

The luminance, L, of the reflected light is defined by

$$L \equiv \delta I / \delta A, \tag{6.26}$$

where δA is the orthogonal projection of δa, as shown in Fig. 6.20. The units of L are then cd m^{-2}. Using equation (6.18) gives,

$$\delta I = \delta(\delta F)/\delta\omega. \tag{6.27}$$

As

$$\delta A = \delta a \cos\theta,$$

then

$$L = \frac{\delta I}{\delta a} \frac{1}{\cos\theta}. \tag{6.28}$$

Fig. 6.20. Diffuse reflection from a surface.

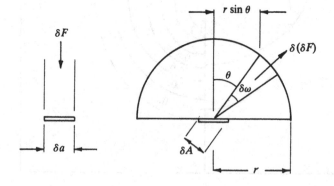

When the surface appears equally bright when viewed from any direction then, from equation (6.26), $L = $ constant, and so,

$$\delta I \propto \cos \theta$$

or

$$\delta I = (\delta I)_0 \cos \theta, \tag{6.29}$$

so that from equation (6.28)

$$L = (\delta I)_0 / \delta a. \tag{6.30}$$

From equation (6.27) the total flux reflected from the area δa is given by,

$$\delta F = \int \delta I \, \delta \omega$$

and from equation (6.29)

$$\delta F = \int (\delta I)_0 \cos \theta \, d\omega.$$

Now,

$$d\omega = 2\pi r \times \sin \theta \times r \times d\theta / r^2$$

giving

$$\delta F = \int_0^{\pi/2} (\delta I)_0 \cos \theta \times 2\pi \times \sin \theta \times d\theta$$

$$= \pi (\delta I)_0.$$

Dividing by δa gives,

$$\frac{\delta F}{\delta a} = \pi \frac{(\delta I)_0}{\delta a},$$

and from equations (6.19) and (6.30)

$$E = \pi L, \tag{6.31}$$

where the units of E are lux and those of L cd m^{-2}.

When the apparent brightness is uniform but only a proportion, r, of the light is reflected, then

$$L = rE/\pi. \tag{6.32}$$

It will be found that writers on photography often adopt the practice of incorporating π into the units of L by putting

$$E = L,$$

so that the units of L, by dimensional analysis, become those of E. They are, however, given a separate name, the apostilb. This is a highly unsatisfactory practice because equation (6.31) applies only to the surface of uniform

Table 6.9. *Typical values of reflectance*

Surface	r
'White-reflectance' paint	0.985
White blotting paper	0.85–0.88
White card, fresh snow	0.8
Light skin (white races)	0.3
'Standard-reference' grey	0.18
Average of typical outdoor scene	0.13
Black paper	0.1
Matt-black paint	0.02–0.03
Black velvet	0.004

brightness and so the factor π cannot be acceptable as a units conversion factor: it often introduces confusion into the writings mentioned.

Typical values of reflectance are given in Table 6.9. Values for objects commonly photographed are surprisingly low.

When an object is lit by a parallel beam of light, such as effectively sunlight, then the mean luminance of that object as viewed from the camera will depend on the relative direction of the light. This is illustrated in Fig. 6.21 where the object is represented by a sphere. The projected area of the lit portion viewed from the camera is the sum of half the circle of the sphere plus the projected area of half the disc which separates the shaded and lit halves of the sphere.

This projected shape of the disc will be reduced by the azimuth angle, θ, from the direction of the camera, and also by the attitude angle, α, of the direction of the light. Geometrical inspection shows that the projected shape will be an ellipse reduced from the circle vertically by $\cos \alpha$ and horizontally by $\cos \theta$. Thus the proportion of illuminated surface viewed by the camera will be

$$\tfrac{1}{2}[1 + \cos \alpha \cos \theta].$$

If the luminance of the shaded area is negligible compared with that of the illuminated area, L_0, and if a mean luminance, $\bar{L}$, is based upon area, then

$$\bar{L}/L_0 = \tfrac{1}{2}[1 + \cos \alpha \cos \theta]. \tag{6.33}$$

This relation is shown plotted in Fig. 6.22. Plotted also are measured values given by Dunn & Wakefield (Ref. [6]) and confirmed by the photographs of Plate 11 of that reference. This shows a reasonably satisfactory agreement with equation (6.33).

So far only the direct illumination from a light source has been discussed but the reflection of the stray light impinging on surrounding surfaces can be of marked significance. This is particularly so when use is made of flash units, partly because the rated output of these units includes a significant factor for the effect of reflection, and partly because not many laboratories would be equipped with instruments to measure the output.

Though the 'mechanical equivalent of light' is 60π lm W^{-1}, integrated over the whole spectrum (Ref. [7]) a typical flash unit only gives about 50 lm W^{-1} (Ref. [6]). When a reflector, of reflectance r, directs the total output, Q lm s within a beam of solid angle ω_B sr then over a flash time, t, the mean output is

$$\overline{It} = rQ/\omega_B \quad \text{cd s.} \tag{6.34}$$

Of the total output rQ, suppose the amount Q_s falls directly on the object and that Q_w falls directly on the surrounding walls. Thus

$$rQ = Q_s + Q_w. \tag{6.35}$$

Fig. 6.21. Sketch of direction of lighting affecting mean luminance.

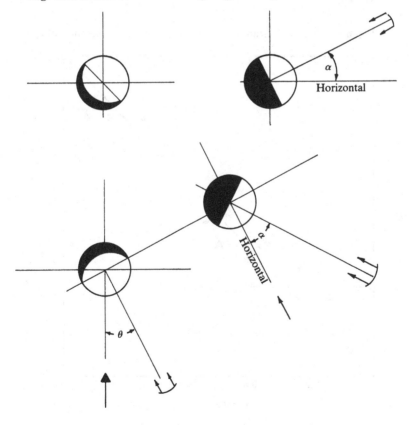

If the object subtends an angle ω_s then

$$\frac{Q_s}{rQ} = \frac{\omega_s}{\omega_B}. \tag{6.36}$$

Denoting the object as surface (1) and the surroundings as surface (2) then the usual relations for radiation give (Ref. [8]), with the view factors F_{12} and F_{21},

$$\begin{aligned}
Q_1^e &= r_1 Q_1^i \\
Q_1^i &= Q_s + F_{21} Q_2^e \\
Q_2^i &= F_{12} Q_1^e + Q_w \\
Q_2^e &= r_2 Q_2^i
\end{aligned} \tag{6.37}$$

Assuming that a fraction, x, of the light reflected on to the object is on surfaces viewed by the camera, then the total light on the object, as viewed by the camera, Q_i, is

$$Q_i = Q_s + x Q_1^i. \tag{6.38}$$

Fig. 6.22. Values of mean luminance as a function of direction of lighting.

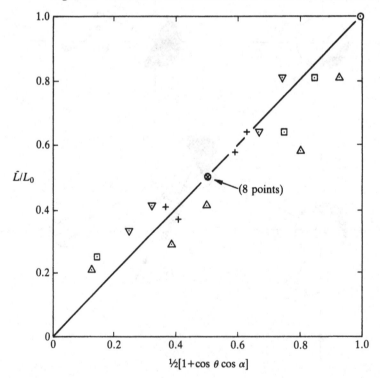

Manipulation of equations (6.35)–(6.38) leads to

$$\frac{Q_i}{rQ} = \frac{(\omega_s/\omega_B)\{1 + x - r_2 F_{21}(r_1 F_{12} + x)\} + x r_2 F_{21}}{1 - r_1 r_2 F_{12} F_{21}}. \tag{6.39}$$

In the case when no light is reflected from the surroundings, then F_{21}, r_2 and x are zero so that equation (6.39) reduces to,

$$\frac{Q_i}{rQ} = \frac{\omega_s}{\omega_B}. \tag{6.40}$$

This is consistent with equations (6.36) and (6.38).

As a numerical example, represent the object as a sphere of diameter, d, in cubic surroundings of side h. Then

$$\frac{A_1}{A_2} = \frac{\pi}{6}\left(\frac{d}{h}\right)^2,$$

whilst $F_{12} = 1.0$ and $F_{21} = A_1/A_2$. For $d/h = \frac{1}{3}$ then

$$F_{21} = 5.82 \times 10^{-2}.$$

Taking $r = 0.7$ for the ceiling and two-thirds of the walls and $r = 0.13$ for the rest gives

$$r_2 = [(1 + 4 \cdot \tfrac{2}{3})0.7 + (1 + 4 \cdot \tfrac{1}{3})0.13]/6$$

$$= 0.48.$$

Then

$$r_2 F_{21} = 2.78 \times 10^{-2}.$$

Let the object subtend a cone angle of 70% of that of the beam angle so that

$$\theta_s/\theta_B = 0.7.$$

With $f = 50$ mm and using 35 mm film from equations (6.1) and (6.5), and with the value of w_f from Table 6.3, then,

$$\omega_s/\omega_B = 0.49.$$

Using $x = 0.5$ and $r_1 = 0.30$ finally equation (6.39) gives

$$\frac{Q_i}{rQ} = 0.75,$$

and so 75% of the output light is incident on the object. Without the surrounding reflection then equation (6.40) gives,

$$\frac{Q_i}{rQ} = 0.49.$$

Now only 50% of the output light is incident on the object; the surroundings have contributed 33% of the total light on the object as viewed at the camera. This seems typical of the allowance made by manufacturers when rating the output of flash units.

6.7   Picture tones

A scene, which is to be photographed, will usually show a range of luminance values from the various parts that are viewed. When the human eye scans such a scene it responds to these various levels whether viewing either the scene or its photographic record.

This response of the eye is such that when it judges there to be a uniform progression of levels of luminance it does not equate these to a sequence of constant change in luminance, δL; rather, it records equal increments of the fractional change, $\delta L/L$ (Ref. [7]). This is the same type of response as that of the ear to changes in level of sound. A common example of this regular progression is seen in the sequence of grey shades from white to black in television test signals; another example is given by Craeybeckx (Ref. [3], p. 59).

This regular progression then corresponds to

$$\frac{\delta L}{L} = K, \tag{6.41}$$

where K is a constant; or, for two successive levels of luminance,

$$L_{n+1} = L_n(1+K).$$

With $K = 1$ the successive values would follow the sequence $1, 2, 4, 8, 16,$ and so on.

In the particular case when each level in a scene occupies equal areas, δa, then equation (6.41) can be expressed as

$$\frac{\delta L}{L} = K \, \delta a,$$

which integrates to

$$\ln L = Ka + \text{constant}. \tag{6.42}$$

This particular gradation of luminance is called a 'uniform tone distribution' and is sketched in Fig. 6.23 where the lowest value of L is arbitrarily scaled to unity. Noting the sequence of tones described, it becomes convenient to write equation (6.42) as

$$\log_2 L = (a/A) \log_2 L_{\mathrm{m}}. \tag{6.43}$$

This distribution represents many scenes containing much detail; examples are given by Dunn & Wakefield (Ref. [6]).

The luminance of an object can be measured by a photographic exposure meter: a photoelectric effect is obtained by facing a suitable surface towards the object and by directing light on to this surface through a suitable lens system so that the luminous flux is accepted from within a finite solid angle. This is sketched in Fig. 6.24.

From an element of area δa of the object, at an angle θ, light of intensity δI and flux δF is captured by the meter area a_m. Then, from equation (6.26)

$$\delta I = L \, \delta a = L \, \delta a \cos \theta$$

and also, from this diagram, δa being effectively a point source,

$$\omega = a_m \cos \theta / (d/\cos \theta)^2.$$

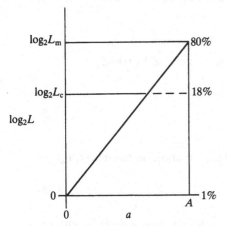

Fig. 6.23. Uniform tone distribution.

Fig. 6.24. Meter acceptance of luminance.

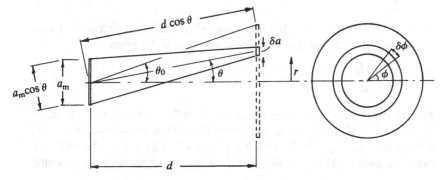

Equation (6.27) is now written as $\delta I = \delta F/\omega$ so that

$$\delta F = L\,\delta a\cos\theta \times a_m \times \cos^3\theta/d^2$$

$$= \frac{La_m}{d^2}\cos^4\theta\,\delta a. \tag{6.44}$$

The sensitive surface in a meter responds to the value of E. Then from equation (6.19) the contribution to this illumination by the flux from δa is δE, where,

$$\delta E = \delta F/a_m$$

$$= \frac{L}{d^2}\cos^4\theta\,\delta a. \tag{6.45}$$

When the view of the meter is a circular cone of angle $2\theta_0$ as shown in Fig. 6.24 then

$$\delta a = r\,\delta\varphi\,\delta r,$$

where φ is the angular coordinate in the plane of a. Further,

$$r = d\tan\theta,$$

so that

$$dr = d\sec^2\theta\,d\theta,$$

and then equation (6.45) gives the total illumination on the meter, E_M, as

$$E_M = \tfrac{1}{2}\int_0^{\theta_0}\int_0^{2\pi} L\sin 2\theta\,d\varphi\,d\theta. \tag{6.46}$$

To perform this integration the distribution of L over a must be known. Taking a mean value, $\bar{L}$, over an annulus, then

$$\int_0^{2\pi} L\,d\varphi = 2\pi\bar{L}.$$

For a scene of fine detail with luminance values uniformly distributed then it can be assumed that $\bar{L}\neq f(\theta)$. Then equation (6.46) gives

$$E_M = \frac{\pi\bar{L}}{2}[1-\cos 2\theta_0]. \tag{6.47}$$

Meters seem to be designed so that the value of θ_0 is about that of the angle of view of a 50 mm lens with 35 mm film. This, from equation (6.5), corresponds to $\theta_0 = 23.4°$ and so $(1-\cos 2\theta_0)/2 = 0.16$. The Weston exposure meter has two scales. Used on the upper scale the acceptance

Table 6.10. *Exposure meter scales*

Make and model	Calibration	Units
Weston IV	$M_w = L$	candles ft^{-2}
	$M_w = L/11.1$	cd m^{-2}
Weston V \ Euromaster	$M_w = \log_2 (40L)$	candles ft^{-2}
	$M_w = \log_2 (3.65L)$	cd m^{-2}
Profisix	$M_w = \log_2 (2.05E)$	lm ft^{-2}
	$M_w = \log_2 (E/5.25)$	lx
Lunasix 3	$M_w = \log_2 (25L)$	cd m^{-2}

angle, $2\theta_0$, is 50° whilst on the 'low-light' scale this angle is 70°. Special versions, called spot-meters, have much smaller angles of view. Commercially available exposure meters are set to read values of either illumination or luminance. Calibration of the scales of four examples are now quoted in Table 6.10; M_w is the numerical scale reading.

It has just been shown that the use of an exposure meter gives a measure of the mean luminance L. This will depend on the distribution of L across the object. With the 'uniform tone distribution' described by equation (6.43) the value of $\bar{L}$ becomes

$$\bar{L} = \frac{1}{A} \int_0^A L \, da. \tag{6.48}$$

From differentiation,

$$\frac{dL}{L} = \ln L_m \, d(a/A),$$

and substitution of this and equation (6.43) into equation (6.48) gives

$$\bar{L} = \int_1^{L_m} \frac{dL}{\ln L_m} = \frac{L_m - 1}{\ln L_m}. \tag{6.49}$$

Tone distributions commonly are not 'uniform'. For example, an object giving a 'uniform tone distribution' but against a white background, would show a distribution as sketched in Fig. 6.25(a); against a black background it would show the distribution of Fig. 6.25(b). If the whole scene contained only the upper level of the previous 'uniform tone distribution' it would appear as in Fig. 6.25(c).

Integrating, for the first two of these cases, equation (6.48), as before, gives in order

$$\bar{L} = \frac{1}{2} \frac{L_m - 1}{\ln L_m} + 0.5 L_m \tag{6.50}$$

and

$$\bar{L}=0.5+\frac{1}{2}\frac{L_m-1}{\ln L_m}. \tag{6.51}$$

As a numerical example consider uniform illumination and for the highest luminance value, $r=0.8$, as corresponding to a matt white surface and $r=0.01$ for the lowest level corresponding to a reflection from black velvet. Scaling this last value to $L=1$ then $L_m=0.8/0.01=80$. Applying these values to the first three cases described gives, in order from equations (6.49), (6.50), (6.51), $L=18$, $L=49$ and $L=9.5$. For the fourth case assume again that L_m corresponds to a value of $r=0.8$ and that the lowest level corresponds to $r=0.08$; then $\bar{L}=3.9$. The four corresponding values of the reflectance for these mean luminance values are 0.18, 0.49, 0.095 and 0.31. The meter then gives four different readings corresponding to four different tone distributions in the objects. But, in all four cases, for correct illumination of the film we require the same luminance from the upper level which is common to all four. If this was used to set the illumination on the

Fig. 6.25. Tone distributions.

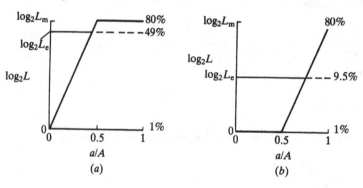

(a)   (b)

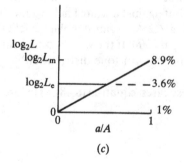

(c)

film then all tones would be equally recorded amongst the four. However, the above values of L are of no help.

What can be done is to set the meter to record only the same tone in each object. In practice this can be done by setting a white card at the object and reading just from that card. Alternatively a meter could be used to read not the luminance from an object but the illumination of that object. Details of techniques depend on the design of the optics of the meter (Ref. [6]). An illustration of the points discussed is given by the diagram of Fig. 6.26 and the accompanying photographs of Fig. 6.27. Fig. 6.26 illustrates a floodlight at position (1) which was above the object so that it shone down at about 45° to illuminate the top and side of the object. At position (2) there was a spotlight to illuminate the front face of the object. Two backgrounds were used, one of white paper and the other of black.

A separate prior experiment enabled the light values given in Table 6.11 to be measured using a 'Weston' exposure meter. Assuming a reflectance, r, of 0.85 for the white paper, then use of the values in this table gives, from equation (6.32), the reflectance of the object as $17.8 \times 0.85/54 = 0.28$ and similarly that of the black paper as 0.02. In this experiment the illumination at the object was measured by the 'Weston' meter with an 'Invercone' attachment. This gives the value tabulated which then corresponds to the luminance of a surface for which $r = 8.9 \times 0.85/54 = 0.14$. This value is used

Fig. 6.26. Photographic and lighting layout.

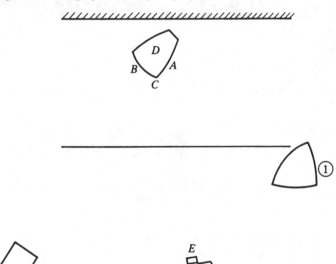

by manufacturers such as 'Weston' and 'Gossen' as representative of the
mean reflectance of the average scene in general photography.

Using a 'Weston' exposure meter the measured luminance values for the
set-up of Fig. 6.26 were obtained as given in Table 6.12. The difference in
luminance values of the object against the two backgrounds then results
from the comparatively large amount of light reflected from the white one.

Fig. 6.27. Illustrative photographs, (*a*), (*b*), (*c*), (*d*).

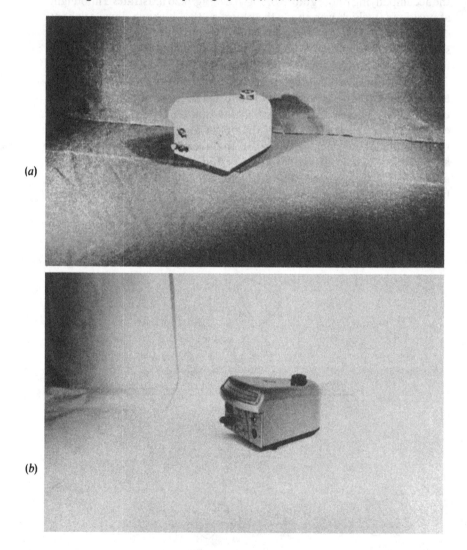

(*a*)

(*b*)

But, further, the values of Table 6.12 show a uniformity of luminance on the three surfaces of the object at *A*, *B* and *D* with the white background which is not obtained with the black one. Separate measurements were taken of white paper set vertically in turn at positions *A* and *B* and the readings are given in Table 6.12. Compared with the separate measurement of *r* for the object of 0.28, now in comparison, and again assuming $r = 0.85$ for the white

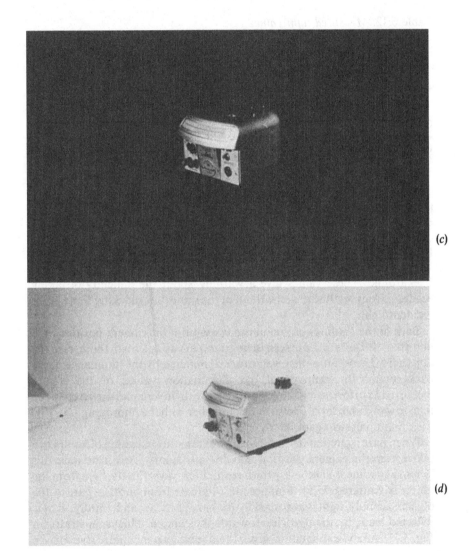

(c)

(d)

Table 6.11. *Measured light values*

Light	Surface	Value	Reflectance
Luminance, cd m^{-2}	White paper	54.0	[0.85]
	Object	17.8	0.28
	Black paper	1.1	0.02
Illumination, cd m^{-2}		8.9	0.14

Table 6.12. *Measured light values*

Light	Position (Fig. 6.24)	Background White Surface		Background Black Surface	
		White paper	Object	White paper	Object
Luminance, cd m^{-2}	A	555	250	280	62
	B	555	250	280	102
	D		250		62
	E		180		10.2
Illumination, cd m^{-2}	C	117		90	

paper, the values of Table 6.12 give $r = 0.38$ for the readings with the white background and $r = 0.19$ and 0.31 for the two readings with a black background. These values give a warning about the difficulty of taking readings from small surfaces without at the same time blocking some of the incident light.

Both of the readings of luminance taken from the camera position at E and given in Table 6.12 are seen to be greatly at variance with those from the object. This is because these values are dominated by the luminance of the background. In comparison, the illumination values of the object, measured as before using a 'Weston' meter with 'Invercone', are given in the final row of Table 6.12. These results, together with the photographs of Fig. 6.27, are discussed again later.

When haze is present in the atmosphere then only a reading of luminance taken from the camera position may be satisfactory. This is because this luminance value is affected by the haze in three ways. Firstly, light from the object is scattered so as to appear to originate from another part of the object; secondly, light is absorbed by the haze particles; and, thirdly, light is reflected back, by the particles, towards the camera. This is illustrated in Fig. 6.28. Analytical solution gives the double exponential curve for the luminance, L, as measured at a distance x from the object (Ref. [9]): the points plotted are of experimental measurements by Dunn and Wakefield (Ref. [6]).

6.8 Exposure flux on the film

Fig. 6.29(a) illustrates the transmission of luminance of a small portion of the object of area δa_1 into illumination of the corresponding area of film, δa_2. The relation of equation (6.44) applies to this effect and is now written as

$$\delta F_1 = \frac{\pi d_s^2}{4u^2} \cos^4 \theta \times L_1 \times \delta a_1,$$

where d_s is the lens stop diameter; and δF_1 is the flux from the area δa_1. On the corresponding area of film, δa_2, the illumination, E_2, is

$$E_2 = \delta F_1 / \delta a_2$$

$$= \frac{L_1}{u^2} \frac{\delta a_1}{\delta a_2} \frac{\pi}{4} d^2 \cos^4 \theta.$$

Inspection of Fig. 6.29(b) shows that by geometrical similarity

$$\frac{\delta a_1}{\delta a_2} = \left(\frac{u}{v}\right)^2. \tag{6.52}$$

Fig. 6.28. Effect of haze upon luminance.

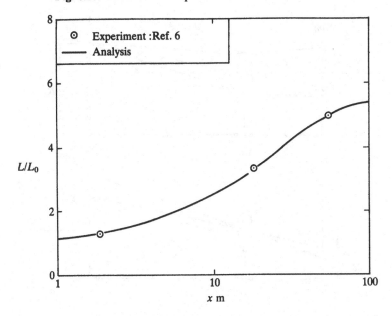

Thus,

$$E_2 = \frac{\pi}{4}\left(\frac{d}{v}\right)^2 L_1 \cos^4 \theta. \tag{6.53}$$

This shows that E_2 is independent of the value of u. It follows that the value of the mean illumination on the film, $\bar{E}_2$, which is defined by

$$\bar{E}_2 a_2 = \int E_2 \, da_2, \tag{6.54}$$

can be evaluated from equation (6.53) for an object whose shape covers a range of values of u. Then using

$$da_1 = dA_1/\cos \theta,$$

$$d\omega = \cos^2 \theta \, dA_1/u^2,$$

and equation (6.53) gives

$$\bar{E}_2 = \frac{\pi}{4}\frac{d^2}{a_2} \int L_1 \cos \theta \, d\omega. \tag{6.55}$$

Fig. 6.29. Diagram showing film illumination.

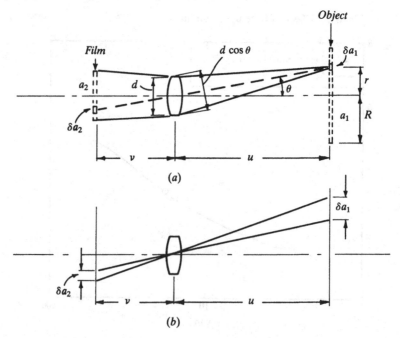

(a)

(b)

It is seen that the value of the integrand is independent of the distribution of u over the object. However, the integration can be performed over an imaginary circular plane of total area a_1, as shown dotted in Fig. 6.29. For this imaginary plane we have

$$\cos \theta = \frac{u}{(r^2 + u^2)^{\frac{1}{2}}},$$

$$da_1 = 2\pi r \, dr,$$

so that equation (6.55) gives

$$E_2 = \frac{\pi^2}{4} \left(\frac{d}{v}\right)^2 \frac{L_1 u^4}{a_1} \int_0^R \frac{d(r^2)}{(r^2 + u^2)^2}, \tag{6.56}$$

where the argument for the introduction of L_1 is the one used previously. This expression integrates to

$$\bar{E} = \frac{\pi}{4} \left(\frac{d}{v}\right)^2 \frac{\bar{L}_1}{a_1} \frac{u^2 \pi R^2}{(R^2 + u^2)}$$

$$= \frac{\pi \bar{L}_1}{4A^2} \frac{\cos^2 \theta_0}{(1 + m)^2}. \tag{6.57}$$

However, the illumination required is that falling on the film which is less than that entering the cone of angle $2\theta_0$. Performing the integration over a rectangle of sides $2w_1$ and $2w_2$ means that equation (6.56) is now

$$E_2 = \frac{\pi}{4A^2} \frac{1}{(1 + m)^2} \frac{u^4}{w_1 w_2} \int_0^{w_1} \int_0^{w_2} \frac{L_1 \, dw_1 \, dw_2}{(u^2 + r^2)^2}.$$

This integrates to

$$E_2 = \frac{\pi \bar{L}_1}{4A^2} \frac{1}{(1 + m)^2} I, \tag{6.58}$$

$$I = \frac{s^2}{2w} \left[\frac{w}{(w^2 + s^2)^{\frac{1}{2}}} \tan^{-1} \frac{1}{(w^2 + s^2)^{\frac{1}{2}}} + \frac{1}{(1 + s^2)^{\frac{1}{2}}} \tan^{-1} \frac{w}{(1 + s^2)^{\frac{1}{2}}} \right], \tag{6.59}$$

where $s = u/w_2$ and $w = w_1/w_2$. As an example, with $w_f = 43.3$ mm and $f = 50$ mm and with $w = 1.5$ then using equation (6.4) with $m = 4.33 \times 10^{-2}$ gives $\tan \theta_0 = 0.4150$. It follows that $s = (1 + w^2)^{\frac{1}{2}} \cot \theta_0 = 4.344$. Then equation (6.59) gives $I = 0.8979$. However, it can now be shown that the simple expression of equation (6.57) is quite adequate provided it is based upon the same area as the above rectangle. This implies an effective cone angle of $2\theta_e$ where

$$\tan \theta_e = \frac{2}{s}\left(\frac{w}{\pi}\right)^{\frac{1}{2}}. \tag{6.60}$$

For the present example this gives $\cos^2 \theta_e = 0.9081$ which is very close to the above value of I.

Also for the present example, equation (6.53) shows that from centre to corner of the film the illumination varies in the ratio of 1:0.73, a 27% drop. For a lens of shorter focal length this effect is even more serious; with $f = 28$ mm the ratio becomes 1:0.42.

Before equations (6.57) or (6.58) can be used it is necessary to relate the value of $\bar{E}_2$ to the response of the film.

The derivation of equation (6.57) assumes that all the flux impinging upon the lens passes through. This is not so. Dunn & Wakefield (Ref. [6]) discuss this point and propose a factor of 0.9 to represent the typical reflection and absorption effects in the lenses of cameras for 35 mm film.

Camera flare is the effect of the reflection from the inside surfaces of the camera of that light entering the lens from outside the angle of view. This light becomes incident upon the film and so adds an increment of illumination that is closely uniform.

If a proportion, p, of all the directly incident light is scattered on the film then this extra illumination is

$$\delta E = \frac{p}{A} \int E_2 \, da_2 = p\bar{E}_2.$$

At any point the total illumination, E_T, is then

$$E_T = E_2 + p\bar{E}_2 \tag{6.61}$$

and the new average value, by integration, is

$$\bar{E}_T = (p+1)\bar{E}_2. \tag{6.62}$$

Examples are given in Table 6.13 for the typical value of $p = 0.06$. In the upper-half of this table some reference levels of illumination quoted by Dunn & Wakefield are given (Ref. [6]).

The first of the two numerical examples is the one considered previously where $L_0 = 1.0$, $L_m = 80$ and $\bar{L} = 18$. It is seen that flare reduces this range of 80:1 down to 39:1. Then, if the value of $\bar{E}_T$ is lined up with the film level of E_D, the lowest level of illumination is close to level E_C and the highest to E_E. The second example considers a case of higher range. The illumination in open shade out-of-doors is from reflection from the surroundings. For an average scene this corresponds to $r = 0.13$ and so gives a factor of $1/0.13 = 8$;

Table 6.13. *Effects of camera internal flare*

			(1) Reference illumination levels
Symbol	Scale level	Reflectance, r	Illumination condition
E_A	1.0	0.006	Slope $= 0.3\gamma$: minimum for reproduction of lowest luminance
E_B	1.2		Safety factor of 20% on E_A
E_C	1.7	0.01	Film speed reference point
E_D	16.9	0.100	Level of $\bar{E}_T$
E_E	80.0		Average high-light
E_F	110.0	0.65	Maximum 'white'
E_G	166.0	0.98	Near maximum diffuse white

	(2) Numerical examples	
Example	I	II
E_0	1.0	1.0
E_m	80.0	640.0
E	18.0	99.0
$\bar{E}_T/E_{0T}$	9.2	15.0
	$E_D/E_C = 10.0$	$E_D/E_B = 15.8$
$E_{mT}/\bar{E}_T$	4.25	6.2
	$E_E/E_D = 4.21$	$E_F/E_D = 5.8$
E_{mT}/E_{0T}	39.0	93.0
	$E_E/E_C = 42.0$	$E_F/E_B = 92.0$

to extend the previous example to cover this lower level makes the range now $8 \times 80 = 640$ but now flare reduces this to 93. This example is shown also in Table 6.13: it is seen to cover the full range of reference levels but with the 20% safety factor noted.

6.9 Film characteristics

If light is transmitted through a negative of uniform 'darkness' then a proportion P is let through. The opacity, O, is defined as

$$O \equiv 1/P,$$

the density, D, is then defined by

$$D \equiv \log_{10} O = -\log_{10} P.$$

For n layers of this film, each layer will pass P of that passed by the layer above and so the total, P_T, is

$$P_T = P^n,$$

and so the total density, D, is

$$D = -n \log_{10} P.$$

This gradation will again give the 'uniform tone distribution' already discussed.

The characteristic for black and white film is typically as sketched in Fig. 6.30 which shows how the density of the negative varies with the exposure H, defined by equation (6.13). In this graph the value of H has been scaled to a reference value H_m which is a measure of the speed of the film. Different forms of light-sensitive layers upon the film will principally shift the curve of Fig. 6.30 to left or right whilst different lengths of development time will principally change the vertical scale of D (Ref. [5]). The definition of the arithmetic film speed, S_A, is for a standardised form of the characteristic curve (Ref. [6]) and is given, for black and white film, by

$$S_A = \frac{0.8}{H_m}, \tag{6.63}$$

with H_m in lx s. This numerical scale of S_A is called both the ASA scale and ISO one. Another numerical logarithmic scale is defined by

$$S_D = 10 \log_{10}(1/H_m). \tag{6.64}$$

This scale is called the Din scale, the ASA-log scale, and the ISO-log scale.

Fig. 6.30. Sketch of black and white film characteristics.

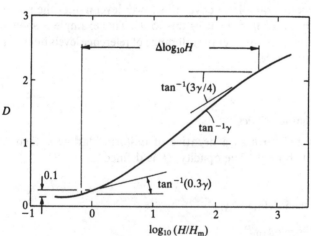

The relation between S_A and S_D is, from equations (6.63) and (6.64),

$$S_D = 10 \log_{10} 2 \cdot \log_2(S_A/0.8).$$

This has been approximated, for the purposes of published tables, by

$$S_D = 3 \log_2 S_A + \text{constant}.$$

Matching the values $S_A = 25$, $S_D = 15$ leads to

$$S_D = 3 \log_2\left(\frac{2^5 S_A}{25}\right) \tag{6.65}$$

as an acceptable approximation.

Ideally, the variation of D across a negative should be proportional to that of the corresponding variation of E: for example, a scene of 'uniform tone distribution' should give the 'uniform tone distribution' of D as described above. As the graph of Fig. 6.30 is a log–log plot then this desirable result would only be obtained if the range of H was contained within the straight-line portion of this characteristic curve. For 'artistic' work satisfactory results are obtained if the lower limit on exposure is at the position where the slope of the curve is 30% of the maximum. Similarly, the upper limit can be set where the slope is 75% of this maximum. The range of exposure, H, is indicated on Fig. 6.30 as the increment $\Delta \log_{10} H$. Values for a range of films (Ref. [5]) are shown plotted in Fig. 6.31 against the film speed, S_D showing a marked decrease in the range with increase in film speed.

The time of development of the negative also affects the range. This is shown in Fig. 6.32 (Ref. [5]). This shows that for nominally the same film the range depends on the format of the film.

Fig. 6.31. Exposure latitude of black and white film as a function of film speed.

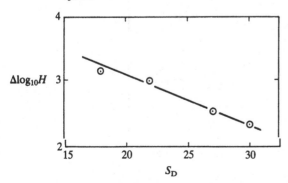

The linear range of the characteristic curve is typically about one-half of the values given and so limitation to this range for scientifically accurate reproduction can be a serious one.

It must be recognised that the speed and exposure range of films are being continually improved by manufacturers. Details of the characteristic curves are obtainable from the better makers.

The first example of Table 6.13 is typical of scientific photography where the subject has the previously quoted range of reflectance and has also a uniform illumination. This requires a range of film illumination of 39:1 or $\Delta \log_{10} H = 1.6$. Inspection of Figs. 6.31 and 6.32 shows that this range is readily available, allowing some latitude on the location of the value of E_T on the film characteristic. For example, referring to the values in Table 6.13 for case I the value of E_{oT} could be lined up with the value of E_A or the value of E_{mT} could be lined up with the value of E_F. The latter setting is disadvantageous in that the film resolution at this higher exposure level can be less and a good print requires more development: however, there can be occasions when this latter setting is required.

The photographs of Fig. 6.27 and the values of luminance and illumination of Table 6.12 may now be studied further.

From Table 6.13 it would be general practice to put $\bar{E}_T/E_c = 10$. However, the object now is of a small range of tone values and so the safety factor on

Fig. 6.32. Exposure latitude of black and white film as a function of development time.

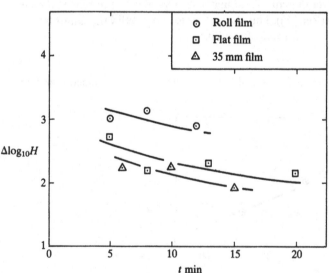

Table 6.14. *Exposure times for Figs. 6.27 and 6.33*

Figure number	L_1 (cd m^{-2})	A	$1/t_e$ (equation (6.66)) (s^{-1})	(camera setting) (s)
6.27(a)	10.2	13.5	1.8	1/2
6.27(b)	180.0	11.0	49.0	1/60
6.27(c)	90.0	11.0	25.0	1/30
6.27(d)	117.0	13.5	21.0	1/30
6.33(a)	204.0	11.0	56.0	1/60
6.33(b)	102.0	11.0	28.0	1/30

the film exposure of 1.2 can be dispensed with so that $\bar{E}_T/E_c = 10/1.2$. From equation (6.13) and referring to Table 6.13 gives $H_m = E_c t_e$. From equation (6.62) $\bar{E}_2 = \bar{E}_T/(1+p)$, and from equation (6.63) $S_A = 0.8/H_m$. Now use of black and white film in tungsten lighting requires the film speed for daylight to be reduced by a factor of 1.25 (Ref. [5]). Combining all these results with equation (6.57) or (6.58) and also (6.59) leads to the relation,

$$\frac{A^2}{S_A t_e} = \frac{1.2\pi}{4 \times 0.8 \times 10 \times 1.25} \frac{1+p}{(1+m)^2} I\bar{L}_i$$

$$= 3 \times 10^{-2}\pi \frac{1+p}{(1+m)^2} I\bar{L}_i. \tag{6.66}$$

Using the values $p = 0.06$, $m = 4.33 \times 10^{-2}$ and $I = 0.8979$ gives exposure times for the photographs of Fig. 6.27 listed in Table 6.14. Referring to this table and also Table 6.12, it is seen that Figs. 6.27(a) and (b), which had the exposure based upon the luminance read from the camera position, are over- and under-exposed respectively, the former being more out. In contrast, Figs. 6.27(c) and (d), which had the exposure based upon the illumination measured at the object, are both at an acceptable level of exposure.

Another example where there is difficulty in assessing the exposure level for the film is illustrated in Fig. 6.33 and also detailed in Table 6.14. It is also an example of use of unusual lighting. The need was to illustrate some glassware containing internal detail; this latter was the small glass nozzle and also the entry point for the side tubing all at the lower portion of the cylindrical containing vessel. With front lighting, as shown in Fig. 6.33(a), detail was lost by reflection from the glass. The exposure of this photograph was determined from a reading of illumination taken from the object position, with a 'Weston' meter plus 'Invercone' giving the value tabulated of 204 cd m^{-2}; this has resulted in a suitably exposed negative. To obtain better information from the photograph, the glass was lit solely by a single floodlight set behind the white paper used as a background. Now the luminance was measured from the camera position and the resulting photograph of Fig. 6.33(b) is seen to be better in showing the detail required. The assessment of the luminance in this case was somewhat fortuitous in that the meter reading was influenced by the luminance of the background.

Fig. 5.39 is an example where scientific requirements can control the printing process. When the negative was printed by a professional photographer the printing exposure was set at 2 s so as to reveal the detail of the background available in the negative. When the printing exposure was extended to 12 s the background was blacked out in this print to give the required result of enhancing the smoke filaments so as to clarify the flow pattern which was the sole object of this photograph.

Continuing the two earlier examples in Tables 6.4, 6.6, 6.7 and 6.8 now applied to case 1 of Table 6.13 gives the results set out in Table 6.15. The final illumination for case (a) of $E_1 = 8.89 \times 10^4$ lx is close to that of bright sunlight: the corresponding value of the flux, $F = 5.25 \times 10^5$, is to be compared with that of 8.0×10^3 for a 'No. 1 Photoflood' lamp (Ref. [6]). If the subject of case (a) is to be lit by two such lamps then one or more of the

Fig. 6.33. (a) Front lighting of glass-work showing unwanted reflections. (b)
(a)   Rear lighting of glass-work revealing internal detail.

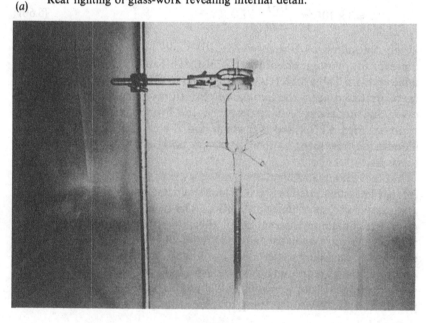

prior conditions must be relaxed. The shutter time has been fixed by the supposed motion of the object: other conditions that could be relaxed are the lens stop setting, A, and the lamp distance d. For example, if a lens of $f =$ 50 mm is used then, from equation (6.10), $A = 4.23$ and the value of $\bar{E}_1 =$ 3.9×10^3 lx. This can be met by setting the two lamps at $d = 2.5$ m. The major effect of this change will be to reduce the size of the image on the film. However, in the calculation of Table 6.15 the nominal value of S_A must now be reduced by the factor of 1.25 if tungsten lighting is used. This then gives the value of F in line 18 and so the lamps would need to be positioned at a distance of 2.25 m.

(b)

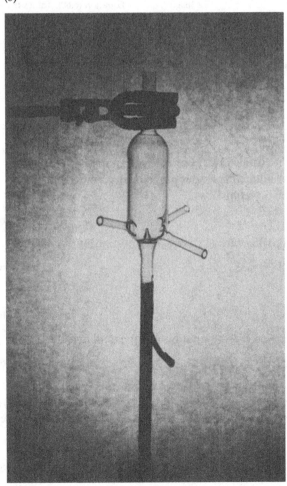

Table 6.15. *Light intensity examples*

Entry No.	Case	(a)	(b)	Source
1	S_D	28.5	30	Equation (6.65), Table 6.6
2	$\Delta \log_{10} H$	2.4	2.3	Fig. 6.31
3	$\log_{10}(E_{mT}/E_{oT})$	1.6	1.6	Table 6.13
4	$\bar{E}_T/E_D$	1.0	1.0	Table 6.13
5	H_m lx s	1.40×10^{-3}	1.0×10^{-3}	Equation (6.63)
6	E_c lx	0.68	4.39	Equation (6.13), Table 6.8
7	$\bar{E}_T = (\bar{E}_T/E_D)$ $\times (E_D/E_c)$ $\times E_c$ lx	6.8	43.9	Table 6.13
8	$\bar{E}_2$ lx	6.4	41.4	Equation (6.62)
9	$\theta_0 = \theta_e$ deg	9.39	16.8	Equation (6.60), Tables 6.3, 4, 7
10	L_1 cd m^{-2}	3.68×10^3	6.92×10^2	Equation (6.57), Tables 6.4, 7
11	$\bar{r}$	0.13	0.13	
12	$\bar{E}_1$ lx	8.89×10^4	1.67×10^4	Equation (6.32)
13	$d = 1.2$ um	2.78	7.75	
14	$I = Ed^2/1.15$ cd	5.97×10^5	8.72×10^5	Fig. 6.19
15	$\omega = 0.28\pi$ sr	0.88		Ref. [6]
16	$\omega(f = 50$ mm) sr		0.52	Equations (6.5), (6.1), Table 6.3
17	F lm	5.25×10^5	4.5×10^5	Equation (6.18)
18	$1.25F$ lm	6.6×10^5		

Referring again to Table 6.15, it was earlier required that the lighting for case (b) was by flash units. The output of such units is indicated by the value of a guide number, G, defined by

$$G \equiv Ad_1. \tag{6.67}$$

A standard (Ref. [10]) relates G to the light output by the relation

$$G^2 = 1.44 \times 10^{-3}\pi I_T S_A, \tag{6.68}$$

where

$$I_T = It \quad \text{cd s.} \tag{6.69}$$

For any lighting arrangement equation (6.24) can be written as

$$\bar{E}_1 = BKI/d_1^2, \tag{6.70}$$

where B is the 'bounce factor' to account for reflection from the surroundings. Use of equations (6.13), (6.32), (6.57), (6.62), (6.63), (6.69) and (6.70) and retaining the value from Table 6.13 of $\bar{E}_T = 10E_c$ gives the result that,

$$I_T = \frac{32}{BK\bar{r}r_f} \frac{(1+m)^2}{1+p} \frac{A^2 d_1^2}{S_A \cos^2 \theta_e}, \tag{6.71}$$

where $\bar{r}$ is the object scene reflectance; and r_f is the reflection factor of the flash light reflector.

The relation between equation (6.68) and equation (6.71) is illustrated by the following example:

For
$$f = 50 \text{ mm}; \quad m \simeq 0; \quad 35 \text{ mm film}; \quad p = 0.06; \quad \bar{r} = 0.16; \quad r_f = 0.9;$$
$$B = 0.75/0.49; \quad K = 0.88;$$
then
$$G = 6.7 \times 10^{-2}(S_A I_T)^{\frac{1}{2}},$$

compared with equation (6.68) which gives
$$G = 6.7 \times 10^{-2}(S_A I_T)^{\frac{1}{2}}.$$

This example then justifies the value given by equation (6.68) but also indicates its limited application.

It should be noted that guide numbers are always quoted with reference to a value of S_A so that, from equation (6.68), it is a measure of I_T that is given.

Returning to the example of case (b) in Table 6.15 with $I = 8.72 \times 10^5$ cd and assuming a flash duration of 2×10^{-4} s then $I_T = 174$ cd s. Equation (6.68) gives $G = 8.88$ and if the flash units available are quoted as $G = 36$ at $S_A = 100$ then this difference can be resolved from equation (6.67) by setting $A = (36/8.88) \times 3.4 = 13.8$. Alternatively the flash units could be moved to a distance of $d_1 = (36/8.88) \times 7.75 = 31.4$ m; clearly altering the lens stop is a more convenient change.

FILM RECIPROCITY

Writing equation (6.13) as
$$H_m = Et, \tag{6.72}$$

the previous analysis has assumed that a film responds to only the value of H and not separately to E and t. This is an approximation that is valid for only a limited range of E: the failure of the relation $E \propto 1/t$ is called the reciprocity failure. Retaining equation (6.72) the relation $H = f(E)$ has been expressed by Kron in the form (Ref. [11])
$$\frac{H_m}{H_{m0}} = \frac{1}{\alpha + \beta} \left[\alpha \left(\frac{E}{E_0} \right)^{\beta} + \beta \left(\frac{E_0}{E} \right)^{\alpha} \right] \tag{6.73}$$

where suffix 0 refers to conditions where H is a minimum and correspondingly $t = t_0$. It is drawn as a logarithmic plot in Fig. 6.34 using the values

$\alpha = 0.5$ and $\beta = 0.2$ suggested by Vaucouleurs *et al.* (Ref. [2]). From equation (6.73) it is found that as $E \to 0$

$$\frac{d(\ln H_m)}{d(\ln E)} \to -\alpha,$$

and as $E \to +\infty$

$$\frac{d(\ln H_m)}{d(\ln E)} \to \beta.$$

The value of t_0 depends upon the film type. For example, Clerc quotes a value of about 1×10^{-2} s for films for general use and of several hours for films made for purposes such as astronomical photography (Ref. [12]).

Fig. 6.34 shows four sets of experimental values for black and white film. Agreement with the Kron equation is seen to be acceptable with the exception of two sets of values at low times. Values of t_0 are listed in this plot: these were determined for a best fit of the data to equation (6.73). The value of t_0 is seen to depend upon the emulsion used. Only one value is in agreement with that just quoted from Clerc. Wakefield has suggested (Ref. [13]) that with black and white film for general use, $2 \times 10^{-3} < t_0 < 1$ s, and the geometric mean of this is 4.5×10^{-2} s.

Fig. 6.34. Reciprocity failure of black and white film.

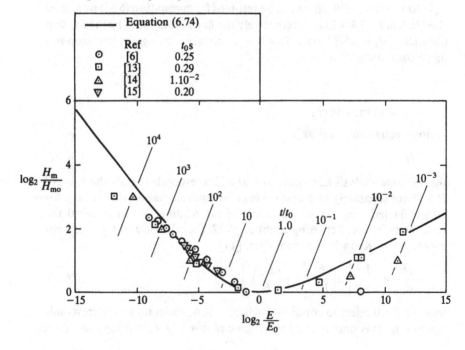

The same reciprocity failure occurs with colour film at low values of E. This is shown in the plot of Fig. 6.35, showing again satisfactory agreement between experimental values (Ref. [16]) and equation (6.73) but now only when $t > t_0$. For small values of t this data shows no reciprocity failure up to the limits shown on the data plots. Values of t_0, shown tabulated in Fig. 6.34, are comparable in order with those for black and white film.

An alternative plot of equation (6.73) is given in Fig. 6.36. From this diagram a change in either t or E, or a combination of both, can readily be determined to account for reciprocity failure.

Using colour film at exposure times that result in reciprocity failure usually also results in a loss of correct colour rendering. This can be corrected by use of filters, as discussed in the next section.

Whilst reciprocity failure is important for very low values of illumination of the film, E, it is also important with very short-time exposures that obtain in flash photography and in very high-speed photography. For the previous example of case (b) with $t = 2.10^{-4}$ s and with a value of $t_0 = 0.2$ s then $t/t_0 = 1 \times 10^{-3}$. From equation (6.73) or Fig. 6.36, $H_m/H_{m0} = 3.70$ and so the value of A or of d would have to be changed so that $Ad_1 = 13.8 \times 7.75/(3.70)^{\frac{1}{2}}$. This condition might be conveniently met by setting $A = 11$ and $d = 5.06$ m.

6.10 Filters for colour film

When the colour of a light matches that of the radiation from a full radiator then the temperature of the radiator, in degrees Kelvin, is called the colour temperature of that light (Ref. [7]). Typical values of light sources are given

Fig. 6.35. Reciprocity failure of colour film.

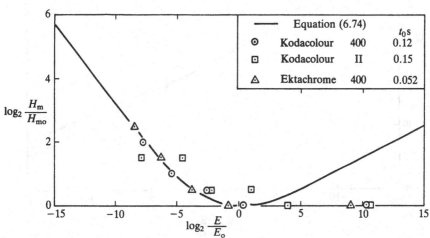

in Table 6.16 (Ref. [16]). Also in this table are the colour temperatures for which the emulsions of daylight and of artificial light films are designed. To balance these various sources with the film requirement colour filters are used. These are obtainable in various densities of either blue or orange.

Table 6.16. *Colour temperatures and mired values*

Source	$T \cdot 10^{-3}$ (K)	M
Blue sky	16.0	62.5
Sunlight (10.00–15.00 hr)	5.8	172.0
Mean noon sunlight	5.4	185.0
Sunlight (16.30 hr)	4.75	211.0
Sunlight (1 hr after sunrise)	3.5	286.0
Electronic flash	6.0	167.0
Fluorescent tube	5.0	200.0
Photo flood	3.4	294.0
Film (daylight)	5.5	179.0
Film (artificial light)	3.2	313.0

Fig. 6.36. Reciprocity correction.

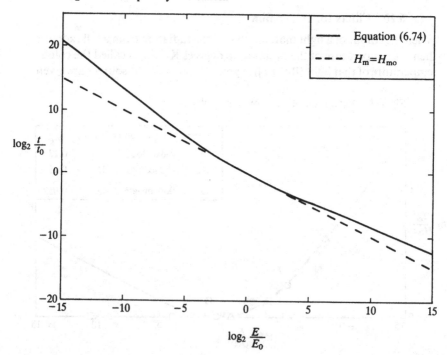

Table 6.17. *Filter for change of colour temperature*

Maker	Reference number	Mired ΔM, value	Filter factor	Colour
Kodak	80 A	−130	4.0	Blue
	80 B	−110	3.2	
	80 C	−80	2.0	
	82 C	−44		
	82 B	−32		
	82 A	−21		
	82	−10		
	81	+8		Orange
	81 A	+18		
	81 B	+27		
	81 C	+35		
	81 EF	+53		
	85 C	+81	1.25	
	85 B	+131	1.6	
Agfa-Gevaert	CTB 16	−190	8.0	Blue
	CTB 12	−140	5.0	
	CTB 8	−87	3.0	
	CTB 4	−45	2.0	
	CTB 2	−31	1.4	
	CTB 1	−15	1.25	
	CTO 1	+7	1.1	Orange
	CTO 2	+18	1.15	
	CTO 4	+49	1.2	
	CTO 8	+95	1.3	
	CTO 12	+150	1.4	
	CTO 16	+190	1.6	
	CTO 20	+220	1.7	
Ilford	810	−165		Blue
	829	−73		
	830	−110		
	831	−57		

If the full radiation of temperature T_1 is passed through a filter to become radiation of temperature T_2 then the quantity $((1/T_2) - (1/T_1))$ is a constant of that filter (Ref. [7]). A mired value of colour temperature, M, is defined by

$$M \equiv 10^6/T \text{ K}^{-1}. \tag{6.74}$$

Then ΔM becomes a filter constant where

$$\Delta M = 10^6((1/T_2) - (1/T_1)). \tag{6.75}$$

Values of M are attached to Table 6.16 and Table 6.17 gives values of ΔM for commercial makes of filter. The sign convention is such that a blue filter has a negative value of ΔM and increases the colour temperature of the light to match the film value: that is, when $\Delta M < 0$ then, from equation (6.74), $\Delta T > 0$: orange filters do the opposite.

The total value of ΔM for two or more filters used together is the sum of the values of ΔM for each filter (Ref. [7]).

For very good colour reproduction the tolerance on the difference of M between light source and film can be ± 5: even ± 10 is usually quite acceptable. Then, the final quality of colour reproduction rests mainly with the quality of film developing and printing processes.

A filter reduces the illumination through it by a ratio known as the filter factor, F: when $F = 2$ only half the light is transmitted and so exposure must be doubled. Published values of F (Refs. [3] and [16]) are shown plotted in Fig. 6.37. From the previous discussion of transmission of light through film emulsions and the summation property of M, it follows that $\log_2 F$ should vary linearly with M. This is shown to be so in Fig. 6.37. The fitted lines are given, for blue filters, by

$$\log_2(F/1.1) = -1.5 \times 10^{-2}M$$

and for orange filters by

$$\log_2(F/1.1) = 3 \times 10^{-3}M.$$

Fig. 6.37. Factors for colour-correction filters.

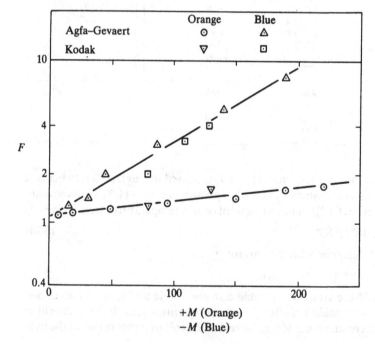

Table 6.18. *Summary of examples*

Case	(a)	(b)
Paper size	B7	B5
Film size	35 mm	120
w_f mm	43.3	101.6
m	4.33×10^{-2}	2.03×10^{-2}
f mm	109.0	130.0
S_A	$\leqslant 570.0$	$\leqslant 800.0$
u mm	2316.0	6462.0
v mm	108.0	130.0
A	$\geqslant 20.1$	$\geqslant 3.4$
t	$\leqslant 2.05 \times 10^{-3}$	$\leqslant 2.28 \times 10^{-4}$
	50 mm	
A	4.23	11.0
d	2.5 m	5.06 m
$G:S_A$		36:100
M (light)	294.0	167.0
M (film)	313.0	179.0
M (film)	179.0	
ΔM (filter)	$-120.0*$	
F	3.9	
A	2.1	

* *Note:* Blue filter. CTB2 and CTB8 or 80 B or 830.

The results of final check calculations for the two examples that have been considered are given in Table 6.18 where the prior results are summarised.

For case (b) the values of M for light and for film just agree to an acceptable tolerance when 'daylight'-type film is used. The same applies to case (a) when 'artificial-light' film is used. The use of 'daylight'-type film for case (a) has a severe disadvantage because a filter of $\Delta M = -120$ is then required. From Table (A1) this could be one of the three possibilities; CTB2 with CTB8, 80B and 830. But the filter factor is then large at 3.9; this could be met by $A = 2.1$ but then with an inability to meet earlier criteria.

6.11 Photographic accuracy

Though the sample calculations given here have been carried through to two- or three-figure accuracy it must be recognised that this is more than is required by the tolerances that are common in photographic apparatus.

For example, a $\pm 10\%$ accuracy might well be found in the settings of shutter speeds, particularly for short times, and in the precision of exposure meters. In this connection the previous discussion of film latitude shows a counteracting tolerance.

Principal notation

A	lens stop number
a, A	area
a_1, a_2	object, film areas
B	bounce factor
c	diameter of circle of confusion
d	stop diameter
d_1	distance of light source
D	density of emulsion
E	illumination
f	focal length
F	luminous flux, filter factor
F_{12}, F_{21}	view factors
G	flash guide number
H, H_m	exposure
I	light intensity
I_T	light output
L	luminance
m	magnification
M_w	exposure meter scale-reading
M	mired value
n	resolution
O	opacity
p	proportion of light
P	proportion of light transmitted
Q	light output
r	proportion of stop reflectance
r, R	radius
s	line spacing
s_i, s_o	image, object sizes
S_A	film speed, arithmetic
S_D	film speed, logarithmic
t	time
t_f, t_e	exposure, shutter time
t_0	zero reciprocity time

T	colour temperature
u	object distance
u', u''	depth of field limits
v	image distance
V	velocity
w_1, w_2	view, film dimensions
w_f	film diagonal size
λ	wave length
θ_0, θ_e	cone angle, effective cone angle
φ	angular coordinate
ω	solid angle

References

[1] Engel, C. E. (ed.) (1968). *Photography for the Scientist*. Academic Press, London.

[2] de Vaucouleurs, G., *et al.* (1956). *Manuel de Photographie Scientifique*. Editions de la Revue d'Optique, Paris.

[3] Craeybeckx, A. H. S. (ed.) (1958). *Gevaert Manual of Photography*. Fountain Press, London.

[4] Keeling, D. (1966). 'Resolving power of b & w films', *Amateur Photographer*, 15 June, p. 824.

[5] Horder, A. (ed.) (1968). *The Ilford Manual of Photography*, 5th Edn (Rev.). Ilford Ltd (England), p. 167.

[6] Dunn, J. F. & Wakefield, G. L. (1981). *Exposure Manual*, 4th Edn. Fountain Press, Watford, England.

[7] Walsh, J. W. T. (1958). *Photometry*, 3rd Edn. Constable, London.

[8] Gibbings, J. C. (1970). *Thermomechanics*. Pergamon, Oxford.

[9] Gibbings, J. C. (1986). 'Effect of haze upon luminance in photography', *Journal of Photographic Science*.

[10] British Standard: 'Methods for determination of photographic flash output and guide numbers', *B.S. 4095: 1970*. British Standards Institution, London.

[11] Kron, E. (1913). *Ann. Phys.*, **4**(41), 751.

[12] Clerc, L. P. (1972). *Photography Theory and Practice* (Ed. D. A. Spencer), Vols. 1 and 2. Focal Press, London.

[13] Wakefield, G. L. (1977). *The Amateur Photographer*. 27 July, p. 119.

[14] *Kodak Handbook for the Professional Photographer* (1973). Professional Publication No. KLP-100, Vol. 1. Kodak Ltd, England.

[15] Maude, M. (1969). *The Amateur Photographer*, 29 Jan., p. 62.

[16] *Kodak Handbook for the Professional Photographer* (1973/1974/1976). Vol. 1 (KLP-100), 1973; Vol. 2 (KLP-200), 1974; Vol. 3 (KLP-300), 1976. Kodak Ltd, England.

7
Interfacing experimental equipment to microcomputers

7.1 Introduction

7.1.1 THE USE OF COMPUTERS IN EXPERIMENTAL WORK

In recent years extensive use has been made of microcomputers coupled to experimental equipment. This was facilitated by the provision of suitable 'interfacing ports' on the microcomputers and by the increased application of analogue devices for the measurement of experimental variables such as pressure transducers, displacement transducers, hot-wire anemometers, thermocouples and accelerometers. Furthermore this development was encouraged by the relative cheapness of modern microcomputers, which with the associated disc drive, printer and monitor, costs only about £1000, in many cases. This means that it is frequently economically viable to dedicate a computer system to a particular experiment. Not only can the computer record the raw data, but it can analyse and display the derived results in one continuing process. It is also possible to save large amounts of data on suitable backing store, whether tape or disc, for further analysis. Finally, it is possible for the computer to control the experiment by transmitting signals from the computer, through the interface ports, to switches or servo-motors, using amplifiers if necessary. This technique can be used, for example, to drive an automatic traverse gear or the stepper-motor of a pressure scanning valve and to switch from one input signal to another.

The purpose of this chapter is to provide a simple introduction to this use of microcomputers and to this end the essential terminology, background theory, hardware description and software techniques are given.

200

7.1.2 TERMINOLOGY

To the uninitiated, computer terminology can be troublesome, especially as so many new words, often formed from acronyms, are involved. In this chapter, every effort is made to explain new terms as they occur and a glossary is provided (Section 7.7) for ease of reference. Electronic computers are relatively new machines, but in their 40 years of development there have been dramatic changes, so that words originally 'coined' to define a particular device or function, may still be used, even though they are now technically incorrect. For example, 'core store' may still be used to describe the computer's main memory, because the original memory devices consisted of small toroidal cores made from ferritic material, whereas most modern main memory devices use semiconductor materials.

To appreciate some of the terminology, a brief outline of the development of computers is helpful. When electronic computers were first created in the 1940s they employed electronic valves along with resistors and capacitors, as did radios, but by the 1950s these valves were replaced by transistors, which were significantly smaller, able to switch at high speed, and consumed much less power than the valve. In 1964 the integrated circuit was invented and this solid-state, electronic circuit was equivalent to a number of transistors and other components all built on a small chip of silicon. Hence the use of the words silicon chip or just 'chip' to mean an integrated circuit (or ic). As the number of transistors per chip increased towards a million, so the terms large-scale integration (LSI) and very large-scale integration (VLSI) were introduced. In 1971 a complete computer was formed on a single silicon chip and this integrated circuit is now called a microprocessor. It does, however, require other chips to support input and output devices and to provide memory for data.

7.1.3 NUMBER SYSTEMS USED BY COMPUTERS

Since electronic switches can have only two states, on or off, it is essential that the number system used by a computer should have two states (0 or 1). This is the binary, or base 2, system. In the general number system to the base n, each increasingly more significant digit has a value n times the previous digit, starting always from unity (n^0) for the least significant digit. Thus in binary ($n = 2$) the column values are:

Most significant digit n^7 n^6 n^5 n^4 n^3 n^2 n^1 n^0 Least significant digit

$$2^7 \quad 2^6 \quad 2^5 \quad 2^4 \quad 2^3 \quad 2^2 \quad 2^1 \quad 2^0$$

$$128 \quad 64 \quad 32 \quad 16 \quad 8 \quad 4 \quad 2 \quad 1$$

The individual binary digits are called 'bits' and eight digits (as shown above) are called a 'byte'. A byte can range in value from 00000000_2 (0_{10}) to 11111111_2 (255_{10}) and these 256 possibilities are sufficient to define all the characters and operators required by the computer. Each binary digit can have the value, or logical state, of 0 or 1, and this logical state is generally transmitted along wires between the various integrated circuits as zero voltage or approximately 5 volts respectively.

The 256 possible states defined by one byte are insufficient for memory address purposes and with most microcomputers two bytes are required to allow 256×256 (65 536) address locations. The more significant byte defines the 'page' of the address and the less significant byte defines the specific memory location on that page.

Since 1024 (2^{10}) is defined as 1 K, the 65 536 memory locations defined by two address bytes are therefore 64 K locations.

The large number of digits involved with binary numbers make mental recall and written or oral transmission difficult. It is usual therefore to employ octal (base 8) or hexadecimal (base 16) for 'human-readable', as opposed to 'machine-readable', data. In octal, each digit can take one of eight values from 0 to 7 and each column value increases by a factor of 8.

In hexadecimal, each digit can take one of 16 values from 0 to 9 and then from A to F. The column values increase by 16. Thus

$$n^2 \quad n^1 \quad n^0 \qquad \text{and} \qquad n^3 \quad n^2 \quad n^1 \quad n^0$$
$$8^2 \quad 8^1 \quad 8^0 \qquad \qquad 16^3 \quad 16^2 \quad 16^1 \quad 16^0$$

$$64 \quad 8 \quad 1 \qquad \qquad 4096 \quad 256 \quad 16 \quad 1$$

Each octal digit is equivalent to three binary digits and each hexadecimal digit is equivalent to four binary digits. Thus one byte can be represented by three octal digits (with one bit to spare) or by two hexadecimal digits. Both number systems are used, octal being employed by some manufacturers (e.g. Intel with their 8080 series of processors), whilst hexadecimal is used by others (e.g. MOS Technology with their 6502 series of processors).

The subscript notation for the base 16 (e.g. $A9_{16}$) is not usually employed. Instead a hexadecimal number is prefixed by an available keyboard character ($ or & are common).

7.1.4 COMPUTER LOGIC AND HARDWARE

Silicon chips used by computers contain electronic gates for the execution of logic operations such as AND, OR, etc. Some understanding of these gates is desirable as they are also used with interfacing circuits. The simplest gate is the inverter, or NOT gate, with one input and one output, which has the opposite state to the input. The symbol for the NOT gate and its truth table are shown in Fig. 7.1.

The AND gate has two, or more, inputs which must *all* be set to logic 1 (i.e. 5 volts) for the single output to be set to logic 1. This is shown in Fig. 7.2, with its truth table. OR, NAND and NOR gates are also shown in Fig. 7.2 and their logic operations should be understandable from the appropriate truth table.

Some commonly used logic integrated circuits are

74LS00   four 2 – input NAND gates (see Fig. 7.3);
74LS02   four 2 – input NOR gates;
74LS04   six inverters;
74LS10   three 3 – input NAND gates;
74LS20   two 4 – input NAND gates;
74LS30   one 8 – input NAND gate.

The heart of a computer is its central processor unit (CPU), in which the processing of the data takes place. The architecture of the CPU varies from one manufacturer to another and is more complex than shown in the simplified schematic diagram, Fig. 7.4. For the present purposes, however, it is not necessary to consider the arrangement and operation of the CPU in great detail. It is sufficient to appreciate that there are registers for storage of data, a control unit which controls the step-by-step operation of the computer and an arithmetic logic unit which performs the operations of addition and subtraction or the logical operations. These registers are connected to the data bus (a group of eight wires) and to the address bus (a group of 16 wires usually) as shown by the broad two-line connections, and serviced by the control unit as shown by the single line connections. For interfacing duties, the ability of the control unit to accept interrupt signals should be noted, as should the usual function of the accumulator to hold one operand before an arithmetic or logic operation and retain the result after the operation.

The address bus, usually consisting of 16 wires and referred to by numbers A0 to A15, communicates primarily with internal devices such as

Fig. 7.1. The NOT gate and the truth table.

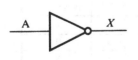

Input	Output
A	$X=NOTA$
0	1
1	0

memory or the keyboard. It also communicates with standard peripherals such as the monitor for the visual display and the printer for the hard copy. But it can also communicate with additional peripheral devices such as the various experimental measurement transducers. For this purpose, each external device is allocated one, or more, addresses and has a suitable logic circuit, the address decoder, to recognise when it has been called. The

Fig. 7.2. AND, OR, NAND and NOR gates.

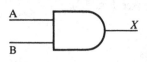

AND gate

Inputs		Output
A	B	A AND B
0	0	0
0	1	0
1	0	0
1	1	1

OR gate

Inputs		Output
A	B	A OR B
0	0	0
0	1	1
1	0	1
1	1	1

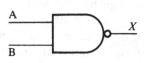

NAND gate

Inputs		Output
A	B	A NAND B
0	0	1
0	1	1
1	0	1
1	1	0

NOR gate

Inputs		Output
A	B	A NOR B
0	0	1
0	1	0
1	0	0
1	1	0

allocation of addresses to memory, keyboard, visual display unit (VDU), printer and other devices is specified for each microcomputer and can be ascertained from a diagram, called the memory map, or from a list of memory map assignments.

The data bus, usually consisting of eight wires (referred to as D0 to D7), is bidirectional, so that data may be sent from the microprocessor (the process of 'writing') or data received (the process of 'reading'). Since eight wires transmit one byte, the resolution is 1 in 255 (the maximum value of the eight-bit byte) and for better resolution it may be necessary to read two

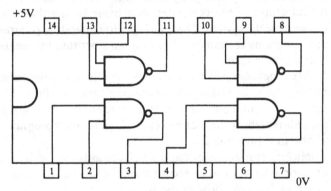

Fig. 7.3. The 74LS00 IC with four 2-input NAND gates.

Fig. 7.4. A simplified schematic diagram of a central processor unit.

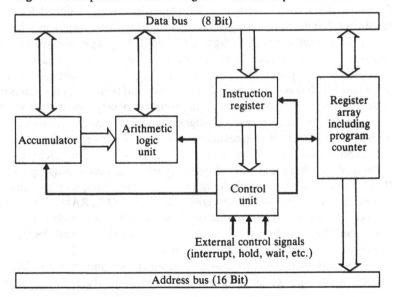

consecutive bytes, even if only 10 or 12 bits are used rather than the full 16. Microcomputers with 16-bit data buses clearly provide more than sufficient resolution for experimental purposes (1 in 65 536).

7.1.5 SOFTWARE

Software is the name given to all the programs, including:

(i)   The operating system programs, which are provided by the manufacturer of the computer to manage the hardware and software of the system. These include disc operating programs and programs to translate the programs, written by the user, into machine code.

(ii) Utility programs, which are also provided by the manufacturer to perform basic tasks, such as making copies of discs.

(iii) Library programs, for a variety of standard operations and procedures which the user can embody in his own program, so saving time and effort.

(iv) Application programs or packages, frequently used in business by people who do not need to have a full knowledge of the program, but only how to use it.

(v)   User programs, which are written by the user for a particular task.

Programs are either written in a high-level language, designed for easy reading and understanding by the user, or in a low-level language, which is a machine-orientated language. High-level-language programs are translated into machine code, so as to be meaningful to the computer, by an operating system program called either a compiler or an interpreter. The compiler, which translates the whole program and then stores the machine-code version, is used when the computer has a large memory and the program does not require many changes. For microcomputers, usually with limited memory size, it is necessary to use an interpreter, which translates and executes each statement in turn, without saving the machine-code version. Virtually all microcomputers support a high-level language called BASIC, which is interpreted. As memory size increases, so many micro-computers now support languages, such as FORTRAN, which are compiled. The majority of microcomputers offer assembly (low-level language) programming, which is slightly less readable, but allows storage of the machine-code version after assembly.

For control and data acquisition purposes, one important difference between an interpreted language such as BASIC and assembly language is the speed of execution. BASIC programs are slow and assembly programs must be used for fast response rates (typically up to 20 kHz).

It is not appropriate to list here all the commands available in the various languages, but attention must be drawn to the two relevant BASIC commands, PEEK and POKE, which are used to communicate with memory or peripheral devices. Digital signals can be sent from the computer to memory or peripheral device by the command POKE I, J where I is the address and J the data. Signals are accepted by the computer using the command $X = \mathrm{PEEK}$ (I) where X is a chosen variable name and I is the address. These two commands are written differently in Acorn BASIC, use being made of the question mark (?) in this case, but the principle is similar.

In assembly language, the corresponding commands to PEEK and POKE are 'load the accumulator' and 'store the accumulator'. For the 6502 series of microprocessors, used by Acorn, Apple and Commodore, these commands are written as the mnemonics LDA and STA respectively.

For further details of the commands available in each programming language, readers are recommended to study the appropriate instruction books provided by the computer manufacturer, especially as there are different 'dialects' to each language (Applesoft and Acornsoft are both variants of standard BASIC).

7.1.6 CHOOSING A MICROCOMPUTER TO SUIT A PARTICULAR APPLICATION

Clearly there are many factors to be considered when choosing a microcomputer for a given task, including the all-important cost factor. The other principal factors are:

size of memory;
program languages accommodated;
input and output ports available;
facilities for backing storage (such as disc store);
speed of data transfer;
data word width.

This must generally imply a detailed study of the specification for each computer and preferably a demonstration of its capabilities. It is, however, possible to consult books and magazines, which list the characteristics of a wide range of computers. Ozanne, 1983 (Ref. [1]), has compiled a useful *Encyclopaedia*, whilst the *A ≡ Z of Personal Computers* by Video Press (Simmons, 1984 (Ref. [2])), gives a convenient comparison of computers with a 1-to-5 star rating of a variety of characteristics, including value for money.

Obviously, the size of memory must be chosen to accommodate the amount of data to be collected and probably the 64 K limit, imposed by a 16-bit address bus, is sufficient. It must be remembered, however, that if more than eight-bit resolution is required, then two memory locations must be used for each datum.

The provision of the BASIC language will satisfy many requirements, but if fast data acquisition or processing is required, it will be essential for assembly language to be available.

As regards input/output ports, it is essential for experimental purposes to have parallel ports available, so that all eight bits can be transmitted simultaneously. Serial ports, with sequential transmission of each bit along one line, would generally be far too slow. Input/output ports will be dealt with in more detail in Section 7.3.

The provision of suitable backing store is essential for saving both the programs and the data. Since cassette tapes are so slow, because they must run through the whole tape to find the particular program or data file (serial access), it is recommended that a disc drive system should be provided. Discs offer direct (or random) access to programs and data files and are significantly quicker than tapes.

The microprocessor of each computer belongs to one of two main groups – those that are register-orientated, i.e. more suitable for control purposes, and those that are memory-orientated and more suitable for data processing. The Intel 8080 series and Zilog Z80 series are examples of the former, whilst the Motorola 6800 series and MOS Technology 6500 series are representative of the latter. Generally the 8080 and Z80 group use high-speed clocks (5 MHz) and many small-step operations whilst the 6800 and 6500 group use low-speed clocks (1–2 MHz) and a small number of powerful operations. Originally, data word widths were eight bits, but 16 and even 32 bit widths are now being used.

Assessing the speed of data acquisition requires a knowledge of the microprocessor clock speed, the number of clock cycles to execute each machine-code instruction and of course whether one word width is sufficient for required resolution. An example, using 6502 assembly language for an Apple ii computer is shown in Fig. 7.5, taken from Shaw *et al.*, 1983 (Ref. [3]). On the right following the semi-colon is an explanation of each instruction and in line 14, for example, LDA $C0D1 is the assembly mnemonic for loading the high byte of a two-byte word from a peripheral device at address $C0D1. This translates into AD D1 C0 in machine code and this instruction takes four clock cycles. The processor clock rate is 1 MHz and thus this particular instruction takes 4 μs. Provided the data is

Fig. 7.5. Apple assembly-language program.

```
0300:                    ORG $0300              1
0300:A9 00               LDA #$0                2   ;LOAD ACC. WITH 0
0302:85 06               STA $06                3   ;STORE IN ADDRESS 06
0304:A9 10               LDA #$10               4   ;LOAD ACC. WITH 10
0306:85 07               STA $07                5   ;STORE IN ADDRESS 07
0308:A0 00               LDY #$0                6   ;LOAD REG. Y WITH 0
030A:48        LOOP      PHA                    7   ;7 CYCLE DELAY TO EVEN TIMES
030B:68                  PLA                    8   ;BY REDUNDANT PUSH-PULL
030C:A9 50     PAGE      LDA #$50               9   ;LOAD GAIN & CHANNEL
030E:8D D0 C0            STA $C0D0             10   ;STORE IN DEVICE ADDRESS C0D0
0311:A5 07               LDA $07               11   ;LOAD PAGE NUMBER
0313:C9 90               CMP #$90              12   ;COMPARE WITH 90
0315:F0 18               BEQ END               13   ;BRANCH TO END
0317:AD D1 C0            LDA $C0D1             14   ;LOAD HIGH BYTE
031A:29 0F               AND #$0F              15   ;STRIP OFF HIGH BITS
031C:EA                  NOP                   16   ;DELAY TO COMPLETE CONVERSION
031D:91 06               STA ($06),Y           17   ;STORE @ ADDRESS @ 07 & 06+Y
031F:AD D0 C0            LDA $C0D0             18   ;LOAD LOW BYTE
0322:C8                  INY                   19   ;INCREMENT Y BY 1
0323:91 06               STA ($06),Y           20   ;STORE @ ADDRESS @ 07 & 06+Y
0325:C8                  INY                   21   ;INCREMENT Y BY 1
0326:C0 00               CPY #$0               22   ;Y=0?
0328:D0 E0               BNE LOOP              23   ;BRANCH TO LOOP IF Y NOT ZERO
032A:E6 07               INC $07               24   ;ADD 1 TO CONTENTS OF ADDRESS 07
032C:4C 0C 03            JMP PAGE              25   ;JUMP TO PAGE
032F:60        END       RTS                   26   ;RETURN FROM SUBROUTINE
```

being stored on one particular page, the program 'loops' at line 23 back to line 7 and, when all the instructions within the loop are counted, 52 clock cycles are involved, resulting in a 'loop' time of 52 μs, or a data acquisition rate of nearly 20 kHz.

One of the main reasons for choosing the Apple II for data acquisition is that it has a distinctive input/output feature. Eight peripheral slots each have access to the address bus, the data bus, and many control lines. Each slot is allocated 16 addresses $C0n0 to $C0nF, where $n = 8 +$ slot number (0 to 7) and the dollar sign ($) indicates a hexadecimal number. Standard integrated circuit 'cards' plug into these slots and permit communication with peripheral devices.

One further feature of microcomputers must be mentioned at this point, although detailed discussion is left to Section 7.2.5. When very rapid data capture is required, a technique called 'direct memory access' (DMA) is used and it should be noted that computers based on the 6502 microprocessor chip cannot readily employ this technique as the address and data buses cannot be disabled separately. Acorn, Apple and Commodore all use the 6502 series of microprocessors.

7.2   Integrated circuits suitable for interfacing to experiments

7.2.1   LOGIC OPERATION GATES

The gates described in Section 7.1.4 are frequently employed, with NAND gates being particularly popular, especially as a NAND gate can be used as an inverter, by connecting all the inputs together, as shown in Fig. 7.11. It is appropriate at this point to explain the standard notation. If an operation is initiated by the line voltage going 'low' (i.e. typically 5 volts, or logic 1, falling to zero volts, or logic 0), then an overbar is placed above the connection name (e.g. $\overline{\text{Read Request}}$). This change of voltage is illustrated beside the line by either ⁵ᵛ⌐L₀ᵥ or ⌐L and referred to as 'not read request', 'not data valid', etc. If there is no overbar then the operation is initiated by a change from zero volts to 5 volts, shown ⌐⌐ and referred to as 'read request', 'data valid', etc.

7.2.2   INTERFACE ADAPTER

As mentioned in Section 7.1.4., data buses are bidirectional. Also the data bus will probably be required to receive signals from several peripheral devices and these signals may only be available for a limited time. Furthermore, the data bus may be required to drive external circuitry which consumes more power than the microprocessor itself can provide. The control pins, address pins and data pins on a 6502 microprocessor are all capable of driving one standard TTL load (10 mW) and some amplification may be necessary. All these factors point to the need for an interface adapter which will control the direction of data transfer, capture (or 'latch') the appropriate signal at the appropriate time and provide power amplification. The integrated circuits, which provide these facilities, are variously known as 'peripheral interface adapters' (PIA), 'versatile interface adapters' (VIA) and 'programmable peripheral interfaces' (PPI).

One interface adapter is the 8255A programmable peripheral interface manufactured by the Intel Corporation. Fig. 7.6 shows a block diagram of

Fig. 7.6. Block diagram of 8255A programmable peripheral interface showing write to control.

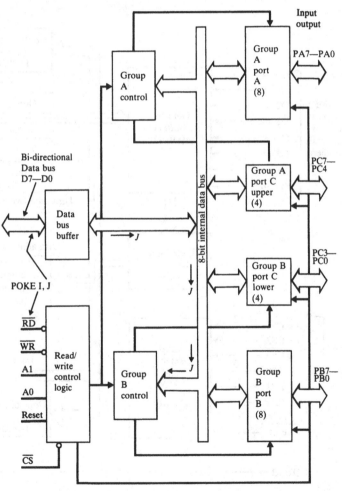

this interface adapter, which essentially has three 8-bit ports for communication with peripheral devices. Port C is split into two four-bit parts and two control units service one eight-bit port (A or B) and one-half of port C. There is also an eight-bit data bus buffer, through which data, control words or status information may be transferred. The read/write and control logic unit manages these transfers, with this particular interface being selected by the 'chip select' line ($\overline{\text{CS}}$) going 'low'. The 'read' line ($\overline{\text{RD}}$) or 'write' line ($\overline{\text{WR}}$) going 'low' selects the direction of data flow, whilst the two

Fig. 7.7. Synertek SY 6520 peripheral interface adapter showing connections and address words.

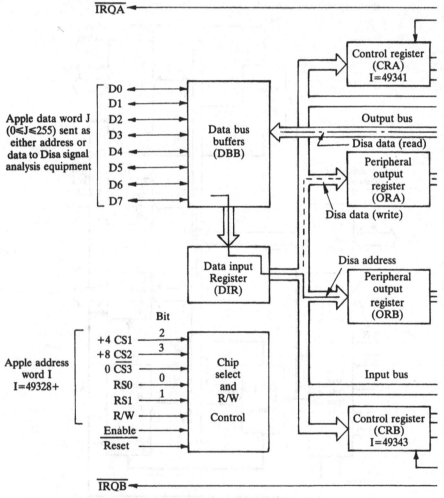

address bits A0 and A1 convey the four possible combinations of logic states (0, 0 or 0, 1 or 1, 0 or 1, 1) to select the control unit or one of the three ports (A, B or C). This diagram also shows a signal being written to the group B control by means of the command POKE I, J where the address I is chosen to set $\overline{\text{CS}}$ 'low' and A0 and A1 'high', the conditions for writing to the control register.

Another interface adapter is the Synertek SY6520 peripheral interface adapter which is shown in use in Fig. 7.7 for connecting an Apple II

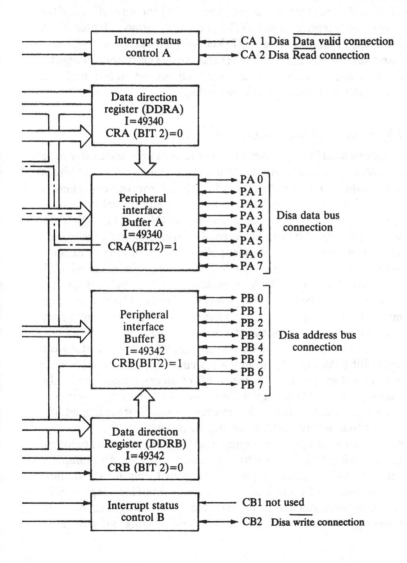

microcomputer to a DISA hot-wire anemometer. This particular application will be discussed in more detail in Section 7.6 but, for the present, it should be noted that it has two 8-bit ports labelled 'peripheral interface buffer' A (and B) and four 'handshake' control lines, CA1, CA2, CB1 and CB2. Like the 8255A adapter described above, it also has a data bus buffer, two control registers and a chip select and R/W control. The architecture of this adapter is shown in slightly more detail than the 8255A with data direction registers shown rather than implied. In this case the respective registers are selected by the logic states present on the register select pins RS0 and RS1 and (except for the control registers) by the state of bit 2 of the appropriate control register. The SY6522 versatile interface adapter which is fully described by Commodore, 1977 (Ref. [4]), is an extended version of the 6520 and has a peripheral control register, an auxiliary control register and two 16-bit counter/timers, which are used on the Acorn analogue interface available from Acorn Computers Ltd.

7.2.3 ANALOGUE/DIGITAL CONVERTERS

Most transducers used for experimental measurement produce an analogue voltage, and servomotors and other external drive devices require a voltage supply and not the digital signal employed by the microprocessor. There is therefore a need for conversion between analogue and digital signals.

One type of analogue-to-digital converter (ADC) is the RS 8703. Conversion is performed by an incremental charge balancing technique. The analogue voltage to be measured is applied, through a suitable resistor, to give an input current of $+10\ \mu A$ maximum, to the I_{IN} pin of the ADC. A reference current I_{REF} of $-20\ \mu A$ is switched in by clock pulses just frequently enough to balance the charge produced in I_{IN}. The total number of I_{REF} pulses to maintain the charge balance is counted and the binary result is latched into the outputs at the end of conversion.

An example of a digital-to-analogue converter (DAC) is the Motorola 1408L8 eight-bit DAC, which through a system of resistors and switches produces a current proportional to the eight-bit binary number set by the logic levels on pins 5–12 of the chip. A CA3140 operational amplifier acts as a current-to-voltage converter, with a feedback resistor providing adjustment of the output voltage for a given digital input.

Some computer manufacturers supply interface systems with analogue-to-digital conversion and other facilities at reasonable cost. An example is the Acorn analogue interface described by Acorn Computers Ltd, 1980 (Ref. [5]). This uses the Synertek SY6522 versatile interface adapter and a 12-bit analogue-to-digital converter capable of selecting one of eight possible analogue inputs. Provided that the controlling software is in machine code then 10 000 conversions can be made in one second or 25 000 conversions per second for eight- or ten-bit resolution. The use of this interface with an Acorn Atom computer has been reported by Shaw *et al.*, 1983 (Ref. [3]).

In some cases interface systems are developed by specialist firms, such as the AI13 analog input system manufactured by Interactive Structures Inc., 1981 (Ref. [6]). The use of this interface with an Apple ii computer for measuring the fluctuating signals associated with turbulent flow is described by Shaw *et al.*, 1983 (Ref. [3]). In this case nearly 20 000 conversions per second were possible at 12-bit resolution because the four most significant bits of the digitised signal are read whilst the remaining eight bits are being evaluated.

Some computers, designed for analogue inputs from 'games paddles', can be used to measure a limited number of analogue voltages. The BBC model B computer, made by Acorn Computers Ltd, is an example. It can measure four analogue voltages in the range 0–1.8 volts and, although a 12-bit converter is used, only ten-bit resolution is guaranteed according to Coll, 1982 (Ref. [7]). Since this represents better than 0.1% resolution, however, it is better than almost all measuring devices and is more than adequate for experimental work. This microcomputer also provides for digital input and output through an eight-bit user port, which also has control lines, all controlled by a 6522 versatile interface adapter.

7.2.4 ENCODERS, DECODERS AND MULTIPLEXERS

In many interfacing circuit arrangements, it is necessary to switch from denary (base 10) numbers to binary numbers and vice versa. Code converters are used for these purposes, and the converter from denary to binary is called an encoder, whilst the converter from binary to denary is called a decoder. In Fig. 7.8 shown in the example in Section 7.4 a 7442A 4–10 decoder is employed to change four input bits to up to ten denary values (only seven are used in the example) to operate ten separate circuits.

Several different devices can be connected to one computer channel through a device called a multiplexer, which scans the peripheral devices to determine which one requires attention.

7.2.5 THE DIRECT-MEMORY-ACCESS (DMA) INTERFACE

As suggested in Section 7.1.6, the data acquisition rate is limited to typically 20 kHz if program-controlled transfer through the processor is employed. However, since it is only necessary to provide an address along with the data for data transfer to memory, integrated circuits called direct-memory-access (DMA) controllers have been developed. These devices have a number of registers, one of which contains the next memory address to be used. The initial address in this register is usually loaded by program, but may also be initialised by hardware. The content of the address register is incremented by one after the transfer of each data byte. Clearly this operation must be halted when all the data has been transferred and to this end a counter register is provided. The counter is set by the program initially, is decremented by one after each transfer, and eventually terminates the DMA operation when the count becomes zero.

DMA controllers usually service four input channels and for this purpose a register called an 'arbiter' is employed to determine which channel has priority. A control register and a bus interface are also needed. DMA controllers can operate in a 'cycle-stealing' mode, when only one or two words are transferred whilst the computer's normal activity is halted for just a few microseconds, but block transfer is more usual for data acquisition purposes.

Two DMA controllers suitable for the 8080 series of microprocessors are Intel's i8257 and the Am 9517 manufactured by Advanced Micro Devices (AMD), whilst Motorola's MC6844 is suitable for the 6800 series of processors. These are described in further detail by Stone, 1983 (Ref. [8]).

Microcomputers are provided with a number of input/output ports, which will be discussed in the next section. For the present, however, it should be noted that pin connections for DMA switching and priority controlling are often included.

7.3   Information transmission between microcomputer and peripherals

7.3.1   SERIAL PORTS

With serial ports, each bit of an eight-bit or 16-bit word is transmitted sequentially along a single line and as a result the data transmission is slow and only suitable for peripheral devices, such as printers, which also operate slowly. Generally, serial ports are not used to control equipment in the immediate vicinity of the microcomputer, but are extremely valuable for communication over long distances. They will be defined here for identification purposes only.

The RS-232-C is a 'recommended standard' of the Electronic Industries Association (EIA) and, although it has 25 lines, many are very specialised and only five lines are normally required. The important lines are 'transmit data', 'receive data', 'request to send', 'data set ready' and the 'signal ground' line.

The 20 mA-current loop standard has two pairs of lines, transmit plus and minus and receive plus and minus. A transmitter is switched so as to send 20 mA-current pulses from a current source to the receiver, which interprets 20 mA as logic 1 and no current as logic 0.

RS-422, RS-423 and RS-449 are other recommended standards for serial ports and further details of all these serial ports are given by Artwick, 1980 (Ref. [9]), Stone, 1983 (Ref. [8]) and the EIA standards.

7.3.2 PARALLEL PORTS

Most microcomputers provide connectors which permit the transmission of eight bits in parallel. These parallel ports are desirable in order to achieve the necessary rate of data transfer, especially with experiments involving transient or fluctuating signals, as with mechanical vibration, electrical oscillation, turbulent flow, etc. It should, however, be noted that one port called the 'parallel printer port' or the 'centronics compatible port' will be dedicated to the printer and not available for experimental use.

The most commonly used parallel port is the IEEE 488 instrument bus defined in the standard adopted by the American Institute of Electrical and Electronic Engineers (IEEE) and based on a handshaking protocol which is patented by Hewlett-Packard, under the name HPIB. This interface has 16 lines, including five interface management lines and three lines for data-byte transfer control (handshaking), as well as the eight-bit data bus. Further details are given by Artwick, 1980 (Ref. [9]), and Stone, 1983 (Ref. [8]). This interface is available on Commodore microcomputers, Research Machines 380Z and 480Z and most Hewlett-Packard machines.

A number of microcomputers provide expansion ports or extension or peripheral slots (Apple ii, iie and iii, Commodore 8032, Dragon, Epson HX20, Hewlett-Packard, Sinclair Spectrum, ACT Sirius) or analogue input ports (Acorn BBC) or user ports (Acorn BBC model B). These ports can be used for interfacing to experiments and the appropriate User Guides should be consulted for further details. For the Apple iie (and iii) the guide gives all the necessary information under the heading 'Expanding the Apple iie' including a table of expansion slot signals for the 50 pins of each slot. These signals include the 16-bit address, the eight-bit data, read/write, interrupt requests, inhibit and the DMA instructions, previously described.

For the Acorn BBC model B microcomputer, the *User Guide* provides the necessary information under the heading 'Analogue input'. The socket labelled 'analogue in' is designed for 'games paddles' or 'joy-sticks' and can also be used to measure up to four voltages in the range 0–1.8 volts. It connects immediately to a μPD7002 analogue-to-digital converter on the motherboard of the computer, already referred to in Section 7.2.3. The eight-bit user port is serviced by port B of a 6522 versatile interface adapter. Port A of the same VIA provides the parallel printer connections. This 6522 VIA is found in the memory map between locations & FE6O and & FE6F.

The computer manufacturer's literature and advertisements should be consulted for full details of interface and expansion sockets and a useful summary can be found in the A ≡ Z published by Video Press, Simmons, 1984 (Ref. [2]).

7.4   Computer control of external equipment

The control of external equipment will almost certainly involve the operation of switches, either to select one of a number of input signals or to switch a power supply greater than available at the output pins of the interface adapter or the I/O port (i.e. one standard TTL load of 10 mW). In this example, shown schematically in Fig. 7.8, the switching is achieved using dual-in-line (DIL) reed relays with 3.7–10.0 volt coils and contacts rated at 0.5 amps, 100 V dc, 10 W. These relays are used to select one of five input channels or two output channels and to select one of three ranges for the analogue-to-digital converter. 2N3704 transistors are connected to the appropriate output pins of port C of the interface adapter or decoder in order to switch 5 volts across the coil of the relay. The circuit for the range relay RR1 is shown in Fig. 7.9. Further details are given by Hardcastle *et al.*, 1982 (Ref. [10]). An example of switching higher power levels is given by Bishop, 1983 (Ref. [11]), who employs a relay with a 12 V coil and contacts rated at 250 V ac, 20 amps for an electric kettle controller. A VN66AF field effect (VMOS) power transistor is employed to switch 12 volts across the relay coil, and this is in turn controlled by the output of a bistable circuit (or flip–flop). The flip–flop is constructed from two NAND gates, cross-coupled as shown in Fig. 7.10, and the flip–flop itself is activated by the computer. The truth table for the flip-flop is shown in Table 7.1. The significance of this flip–flop is that momentary changes to logic 0 in either line A or B cause a change of logic at C, which is retained (or latched) when the lines A and B revert to logic 1.

Switching with power amplification and high voltage relays would be required when controlling servomotors for driving external equipment such as a traverse gear (Hardcastle *et al.*, 1982 (Ref. [10])) or a pressure scanning valve, and for controlling heater elements for temperature regulation.

219

Fig. 7.8. Schematic diagram for interface between Commodore Pet and transducers for air velocity measurement.

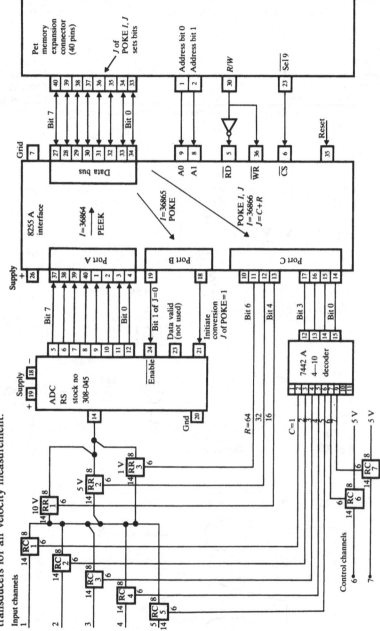

Table 7.1. *The truth table for a flip-flop*

	A	B	C	D	
Initial state	1	1	1	0	Assumed
Drop B to logic 0	1	0	0	1	C changes
Raise B back to logic 1	1	1	0	1	No change
Drop A to logic 0	0	1	1	0	C changes
Raise A back to logic 1	1	1	1	0	No change

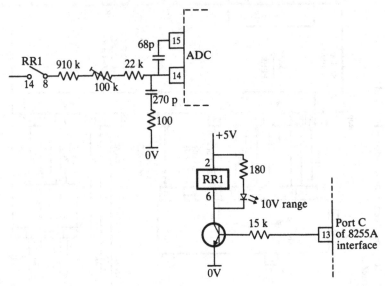

Fig. 7.9. Circuit for range relay RR 1.

Fig. 7.10. A flip-flop.

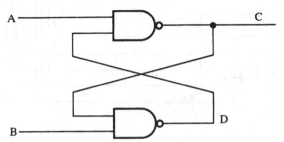

7.5 Analogue-to-digital conversion of transducer outputs

A number of transducer devices are described in Chapter 5 including tachogenerators, linear variable differential transformers, electrical resistance strain gauges, piezoelectric accelerometers, load cells, pressure transducers, hot-wire anemometers and thermocouples. The analogue voltage outputs of these devices range from the order of 100 volts for a tachogenerator, through 5–10 volts for a pressure transducer to 10–50 mV for a thermocouple. Thus if the same analogue-to-digital converter is used for a mixture of input signals it is important to accommodate these different voltages, especially if eight-bit conversion is employed, for otherwise the 1-in-255 resolution at full scale is soon reduced to an unacceptable level. Fortunately, most of the analogue-to-digital converters described in Section 7.2.3 can readily accommodate such voltage variations. The RS 8703 ADC accepts an input current of + 10 μA maximum on pin 14. Using Ohm's law, the resistor to be placed in the line to pin 14 can easily be calculated so that whatever the maximum voltage the charging current is limited to + 10 μA. The circuit shown in Fig. 7.9 has 910 k in series with a 100 k variable and 22 k resistors, giving an adjustable 1000 k to accommodate a 10 volt input. Alternative circuits with 500 k and 100 k resistors, chosen by the 'range relays' RR2 and RR3 (Fig. 7.8), provided 5 volt and 1 volt ranges.

The AI13 analog input system also described in Section 7.2.3 is extremely versatile. Not only can 16 separate input channels be selected, but eight different voltage ranges can be chosen (from 0 to 100 mV through to $\pm$ 5 V). Also, since 12-bit conversion is employed, the full scale resolution is 1 in 4096 (or 0.024%). It also follows that two read operations are required to store the result in the memory of an eight-bit microcomputer. The first operation reads the most significant four bits (or nybble) from the chosen address of the device plus one, whilst the second operation reads the least significant eight bits (or byte) from the device address. Shaw *et al.*, 1983 (Ref. [3]) describe the use of this analog input system in more detail.

The Acorn analogue interface does not provide range selection, but, since 12-bit conversion is employed, a resolution of 2.5 mV is achieved for a voltage range from − 5.12 V to + 5.11 V. A voltage-divider circuit would be required to accommodate higher voltages and some amplification of the input signal needed to deal with voltages in the mV range.

When using analogue input sockets, which are provided for games paddles or joysticks, the manufacturer's guide must be consulted to ascertain the voltage range and resolution. The BBC microcomputer accepts voltages up to 1.8 V with a guaranteed ten-bit resolution. Again it will be necessary to use a voltage divider for higher voltages.

7.6 Examples of interfacing to experiments

The aim of this chapter has been to describe the general principles of interfacing to experiments with only limited reference to specific examples. In the last five years or so, there have, however, been many papers describing particular applications, principally in spectrometry, but also including calorimetry, optical analysis, laser Doppler anemometry, vehicle performance, waveform recording, low-frequency friction studies, photon counting, recording of video data and of seismic events, and control of furnace temperature and of filter fabrication. 8080, Z80 and 6802 microprocessors have been used and Commodore PET, AIM 65 and Apple II microcomputers have been employed most frequently. Acorn System 1 and Research Machines 380Z microcomputers and LSI-2 and PDP-11 minicomputers have also been used.

More than 25 examples of computer interfacing to experiments have been reported in the *Journal of Physics E. Scientific Instruments* in the period 1979–84 and many more in the *Review of Scientific Instruments*. Although all these papers are not given in the list of references at the end of the chapter, readers are referred to the index of papers in both these journals under the subheading '06.50 data handling and computation'.

One specific example has already been illustrated by Fig. 7.8, which shows the schematic diagram for interfacing a Commodore PET model 3016 to an experiment involving the measurement of the velocity distribution across an air jet. A pitot tube for measuring the velocity is moved across the jet by a traverse gear driven by a servomotor controlled by the control channels 6 and 7. The pneumatic pressure difference between the pitot tube and the local static pressure is applied to a pressure transducer and the resultant analogue voltage fed to input channel 1. The position of the pitot tube is detected by a recording potentiometer, with the total voltage across the potentiometer applied to input channel 3 and the voltage split, which varies with the position, fed to input channel 2. The computer program, written in BASIC and using the POKE I, J command, is used first to set the input/output mode of the interface. Data is then supplied from the calibration of both the pressure transducer and the recording potentiometer and then further POKE I, J commands activate the appropriate relays to switch each signal in turn through the correct range channel to the analogue-to-digital converter. A further POKE I, J command, sent through port B of the interface, initiates conversion by the ADC and then a PEEK (I) command reads the latched output of the ADC through port A of the interface and into computer memory. The chip select $\overline{CS}$ is connected to the select line Sel 9 of the computer and this fixes the base address I as \$9000 (or 36864_{10}) for port A. Port B, port C and the interface itself have addresses incremented by 1, 2 and 3 respectively from this value. The value of J, transmitted via port C, is determined from $J = C + R$ where C is the channel number and R is 16, 32 or 64 for the 10 V, 5 V or 1 V range respectively. These values are defined by the data bit connections to port C, where, for example, bit 4 (valued at 16) is connected to the 10 V range relay RR1. For port B, the initiate conversion pin for the ADC is connected to the port pin 18, which transmits bit 0 of the data word. For this bit set high on its own the value of J is 1. Further details are given by Shaw *et al.*, 1981 (Ref. [12]), and by Hardcastle *et al.*, 1982 (Ref. [10]).

A second example is illustrated in Figs. 7.11 and 7.12 which show a block diagram and a wiring diagram for the interface between a hot-wire anemometer system (made by DISA) and an Apple II microcomputer. In this case the DISA anemometer is itself coupled to signal analysis equipment (SAE) which provides a digital output of the measured mean velocity and also rms and correlation values. The SAE has an interface connector with an eight-bit data bus connected as shown to port A of the 6520 peripheral interface adapter (PIA) and an 11-bit address bus, eight lines of which are connected to port B of the PIA. Fortunately the other three address lines can be set to logic 0 (A5 and A10) or logic 1 (A6) in this particular application as shown in Fig. 7.12. Read and write requests, address enable and data valid connections are made to three of the four handshake lines (CA1, CA2 and CB2) of the PIA. The connections between the PIA and slot 3 of the Apple II computer include an eight-bit data bus, a read/write request line ($R/\overline{W}$), a device select line ($\overline{CS3}$), two register select lines (RS0 and RS1) and two chip select lines (CS1 and CS2). Again POKE I, J commands allow the signals to be sent to the various registers of the 6520 PIA, as shown in Fig. 7.7, and onwards to the signal analysis equipment (SAE) as required. The SAE itself has an I/O buffer multiplexer which accepts signals for device selection, channel and time constant and transmits the data. The address word sent via port B of the PIA, selects the type of device (mean or rms) and each of the decimal digits of the converted data, one-by-one (select output data).

Since slot 3 of the Apple II was employed, the base address is $COB0$ (49328_{10}) as shown in Fig. 7.7, which also shows the addresses of the individual registers within the peripheral interface adapter. Further details of this system are given by Hardcastle *et al.*, 1983 (Refs. [13] and [14]). As a final example, Laker *et al.*, 1982 (Ref. [15]), describe the interfacing of a hot-wire anemometer to an Apple II computer with direct memory access controlled by a Motorola MC 6844 DMA controller chip. Sampling rates are programmable between 122 Hz and 125 kHz for single channel operation.

224

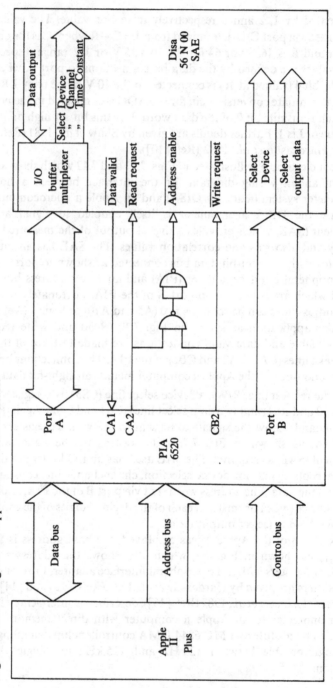

Fig. 7.11. Block diagram for Apple–anemometer interface.

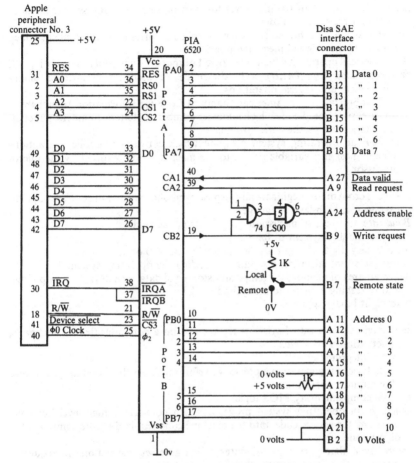

Fig. 7.12. Wiring diagram for Apple–anemometer interface. (*Note:* the overbar on a connection name (e.g. Read Request) indicates that the operation is initiated by the line voltage going 'low' (i.e. typically 5 volts, or logic 1, falling to zero volts, or logic 0.) This is referred to as 'not read request'.

7.7 Glossary

access time: the time taken to read a word of information from computer memory.

accumulator: a microcomputer component, usually called a register, in which arithmetical and logical data are stored and manipulated.

acronym: a word formed from the set of initial letters of other words.

ADC: abbreviation for analogue-to-digital conversion.

address: a numerical value that locates the position in memory of a particular word of information.

address bus: the set of parallel wires for transmitting the address of a memory location or a peripheral device.

analogue signal: a continuously varying voltage, current or other signal which represents the original measured quantity.

AND: a logical operation which gives logic 1 output only when *all* inputs are logic 1.

arithmetic logic unit (ALU): a hardware device used for performing binary arithmetical and logical operations.

ASCII: abbreviation for 'American Standard Code for Information Interchange'. An extensively used code which allocates numbers to letters, numbers, punctuation marks and mathematic symbols.

assembler: an operating system program which converts symbolic instructions (mnemonics) and variable names to the machine code required by the microprocessor.

BASIC: acronym for 'beginners all-purpose symbolic instruction code'. A popular, easy-to-use, high-level language.

Baud rate: the data transmission rate in bits per second.

binary: a number system to base 2.

bit: a binary digit, having only two possible values, 0 or 1.

bootstrap: an initialisation program for loading the operating system.

buffer: an area of memory for temporary storage of data, or a device for amplifying and refreshing a signal to a required level.

byte: eight binary digits.

character: one of a set of symbols represented on a computer (letters, numerals, punctuation marks, etc.).

chip: an integrated circuit.

clock: a device for generating a precise waveform which is then used to synchronise all computer operations.

CMOS: complementary MOS logic.

compiler: an operating system program which converts a high-level language program, the source code, into the machine code required by the computer, the object code.

control unit: that part of a computer which accesses instructions in sequence, interprets them and controls their implementation.

core: commonly used to describe memory, which originally consisted of ferritic rings which could be magnetised by current flow through interlacing wires.

CPU: abbreviation for central processor unit consisting of a number of integrated circuits in which the processing of data takes place.

DAC: abbreviation for digital to analogue conversion.

data bus: the set of parallel wires for the transmission of data.

disc: a flat, circular, metal, or plastic device coated with magnetic oxide and used for storing data.

DMA: abbreviation for 'direct memory access', which is the process of by-passing the microprocessor and transferring data directly to or from memory.

dynamic memory: memory in which the states are stored as charges, which require periodic restoration.

EOF: the 'end-of-file' character.

EPROM: acronym for erasable programmable read-only memory. EPROMs may be erased by ultra-violet light.

fetch: the act of retrieving an instruction or data from memory.
file: a block of data with a defined format.
flag: a one-bit status signal held in memory until reset.
flip-flop: a one-bit memory, with two states.

handshaking: the process of signalling between computer and peripherals (e.g. data ready for transfer and data received).
hard copy: a record of computer output printed on paper or other permanent material.
hardware: all the physical devices and material of a computer system.
hexadecimal: a number system to base 16 (digits 0 to 9 and letters A to F).
high-level language: an application-orientated program language, convenient for users, and which must be translated to machine code before use by an operating system program.

integrated circuit: a solid-state electronic circuit equivalent to a number of transistors and other components built in the same package on a base of silicon.
interpreter: an operating-system program which translates high-level language statements into machine code and executes them one at a time. For micro-computers with limited memory, this avoids storage of the compiled machine code version of the program.
interrupt: the act of diverting the processor from the execution of the current program to a more urgent task.

joystick: an electromechanical device consisting of a rigid stick, which can be tilted in any direction to create electrical signals.

K: a unit of memory size equivalent to 1024 (i.e. 2^{10}).

logic gate: an electronic circuit designed to perform logic operations such as AND, OR, etc.
logic state: the signal state defined as logic level 1 when a voltage of approximately 5 volts exists (typically for TTL logic) and logic level 0 when the voltage is approximately zero.
low-level language: a machine-orientated language in which each program instruction is similar in representation to a single machine language instruction.
LSB: abbreviation for 'least significant bit'.

machine code: a computer program in binary form.
memory map: a diagram indicating all the locations in memory and the assignment of each location to specific tasks, including operating system usage.
memory-mapped I/O: the assignment of memory addresses to the input and output ports, as well as memory, so as to allow data transfer with peripheral devices by reading from and writing to those addresses.
microcomputer: a fully operational computer built around a microprocessor chip.
microprocessor: a single integrated circuit that contains the control and arithmetic logic unit of a computer.

mnemonic: the name given to an abbreviated computer instruction in assembly-language programming.

MOS: metal-oxide silicon device.

MSB: abbreviation for 'most significant bit'.

multiplex: the process of transmitting several signals to one device, which is capable of discrimination between those signals.

NOT: a logical operation with an output of reverse logical state to the input.

nybble: half a byte (4 bits).

object code: the machine-code translation of a program after compilation or assembly. Object code may be loaded and run directly.

octal: a number system to base 8.

operating system: a program or set of programs provided by the computer manufacturer for managing the hardware and other software of a computer system.

OR: a logical operation which gives logic 1 output if *any* one of the inputs is logic 1.

overflow: the act of exceeding the numerical capacity of the arithmetic logic unit.

page: a defined number of memory locations, typically 256 locations corresponding to the number of values of an eight-bit word.

parity check: a method of checking the correctness of data by assessing the number of bits set at logic 1 in a byte or word and comparing with either the even or odd number preselected for the system.

PEEK: a command in BASIC for inspecting memory.

peripheral: an external device connected to a computer.

phase: the displacement between waveforms of a multiple clock system.

PIA: abbreviation for 'peripheral interface adapter'. An integrated circuit with input/output ports, 'handshaking' control lines and buffers to facilitate the connections between computer and peripheral device.

pixel: a picture element.

POKE: a command in BASIC for setting the contents of memory.

port: the connection, containing control logic and data storage, between computer and peripheral device.

program-controlled input/output: the input- or output-transfer of data under the control of a program executed by the microprocessor.

program counter: a processor register containing the address of the next instruction to be executed.

PROM: acronym for 'programmable read-only memory'.

quad-gate package: an integrated circuit with four gates on the one chip.

random access: the method of accessing any memory or storage location directly, without passing sequentially through other locations.

random access memory (RAM): memory with direct access, usually with both read and write capability.

read only memory (ROM): memory which can be accessed, but not altered.

real-time system: a system which is capable of receiving continuously changing data from external sources and process that data sufficiently quickly to control or influence the sources of data.

register: a high-speed memory, usually internal to a processor or I/O port.

sector: an area of a disc defined by angular limits.
semiconductor memory: memory based on semiconductor materials and capable of storing signals as small electrical charges.
serial access: the method of accessing data which requires all previous items of data to be read first.
serial data transfer: data transmission on a sequential bit-by-bit basis.
sign bit: a bit used to indicate the sign of a number. Usually the most significant bit, set to 0 for positive or 1 for negative.
software: the programs, including operating system and user programs.
source code: the readable input program in assembler or high-level language before translation to machine code.
static memory: memory capable of retaining its contents without periodic refreshing.
status: the condition or state of a device.

track: a concentric ring on a disc.
TTL: abbreviation for 'transistor–transistor logic'.

vectors: addresses reserved as jump addresses to be used whenever interrupts, traps or other program actions require.
VMOS: vertical MOS field effect transistor.

word: a defined number of bits used for one item of data. Typically eight bits (one byte) or 16 bits (two bytes).

References

[1] Ozanne, K. (1983). *The Interface Computer Encyclopaedia.* Interface Publications, London.
[2] Simmons, Teresa (ed.) (1984). $A \equiv Z$ *of Personal Computers.* Video Press, London.
[3] Shaw, R., Hardcastle, J. A., Riley, S. & Roberts, C. C. (1983). 'Recording and analysis of fluctuating signals using a microcomputer', *BHRA International Conference on the Use of Micros in Fluid Engineering*, BHRA Fluid Engineering, Cranfield, pp. 185–96. To be republished in *Measurement* by The Institute of Measurement and Control for IMEKO, 1985.
[4] Commodore Business Machines (1977). *MCS6522 Versatile Interface Adapter.* Commodore Buromaschinen GmbH, Neu-Isenburg.
[5] Acorn Computers Ltd (Oct. 1980). *Acorn Technical Manual. Acorn Analogue Interface* (Issue 1). Acorn Computers Ltd, Cambridge.
[6] Interactive Structures Inc. (1981). *AI13 12-Bit Analog–Input System for Apple. Users' Manual.* Interactive Structures Inc, PO Box 404, Bala Cynwyd, Pennsylvania.
[7] Coll, J. (1982). *BBC Microcomputer System User Guide.* British Broadcasting Corporation, London.
[8] Stone, H. S. (1983). *Microcomputer Interfacing.* Addison-Wesley Publishing Co., Massachusetts.

[9] Artwick, B. A. (1980). *Microcomputer Interfacing*. Prentice-Hall Inc., New Jersey.

[10] Hardcastle, J. A., Shaw, R. & Kamal Ariffin, A. (1982). 'Control of an experiment and analysis of results by minicomputer', *J. Phys. E.: Sci. Instrum.*, **15**, 44–6.

[11] Bishop, O. (1983). *Simple Interfacing Projects*. Granada Publishing Ltd., London.

[12] Shaw, R., Hardcastle, J. A., Jusoh, A. R. B. & Bongkik, M. J. (1981). 'Experimental data recording and analysis using a PET minicomputer', *J. Phys. E.: Sci. Instrum.*, **14**, 301–5.

[13] Hardcastle, J. A., Shaw, R. & Shigemi, M. (1983). 'Interfacing digitised DISA-anemometer output to an Apple II computer', *BHRA International Conference on the Use of Micros in Fluid Engineering*. BHRA Fluid Engineering, Cranfield, pp. 315–26.

[14] Hardcastle, J. A., Shaw, R. & Shigemi, M. (Sept. 1983). 'Interfacing digitised DISA-anemometer output to an Apple II computer', *J. Microcomputers and Microsystems.*, **7**, 314–19.

[15] Laker, J. R., Verripoulos, C. A. & Vlachos, N. S. (1982). *A Microprocessor Controlled Data Acquisition and Processing System for Hot-wire Anemometry*. Department of Mechanical Engineering, Imperial College, London.

Bibliography

Bradley, R. J. (1983). *Computer Studies*. Charles Letts and Co. Ltd, London.

Cluley, J. C. (1975). *Computer Interfacing and On-line Operation*. Crane, Russak and Co. Inc., New York.

De Jong, M. L. (1980). *Programming and Interfacing the 6502 with Experiments*. Howard W. Sams and Co. Inc., Indianapolis.

8
Errors in experimentation

Every measurement involves an error. The nature of errors may vary and so may their magnitudes but total elimination of errors from experimentation and testing remains beyond human power.

Irrespective of its formulation or nature, error is the deviation of a measured value from the corresponding true value. Errors comprise a number of constituent components. Some of them can be eliminated completely by special care and attention, and some cannot.†

Because errors cannot be avoided in general one must learn how to live with them. One must also be able to assess their magnitude and, moreover, to be able to control their magnitude according to justified needs.

Errors are frequently random or include a random component. Therefore statistical methods and the probability theory play an important part in the analysis of errors. To be able to predict the effects of errors upon measurements and/or processes is often absolutely vital. Engineers, it is sometimes said, are on average less error conscious than physicists. If this is true, there is obviously a cause for concern.

The relevant literature on experimental errors is extensive and it would be difficult to list it in detail in this chapter. It may be, however, worth noting that the following textbooks: Beckwith & Buck (Ref. [1]), Brinkworth (Ref. [4]), Cook & Rabinowicz (Ref. [5]), Richards (Ref. [8]), could be valuable as supporting reading. Further details on the analytical

---

† There is one exception to this, namely counting, which is often classified as the simplest form of measurement. Counting of numbers, especially of small numbers, can be executed without any error at all.

231

treatment of data can be found in: Pugh & Winslow (Ref. [7]), Beers (Ref. [2]), Moroney (Ref. [6]), and Braddick (Ref. [3]).

8.1   Some useful definitions and concepts of error quantities

ABSOLUTE ERROR

The *absolute error E* is equal to the actual difference between the true value X_T and the corresponding measurement (observation) X_M, or calculation, if such a calculation involves measured quantities or/and approximations.

Absolute error $E = X_T - X_M$.

RELATIVE ERROR

A non-dimensional form is obtained by dividing the absolute error by its measured value. Such a form is known as the *relative error*.

$$\text{Relative error } e = \frac{E}{X_M} = \frac{X_T - X_M}{X_M} = \frac{X_T}{X_M} - 1.$$

TOLERANCE

The *error tolerance* is the region within which the true value can be found. If δ_{max} and δ_{min} denote the maximum and minimum error limits then

$$X_T = X_{M-\delta_{min}}^{+\delta_{max}}$$

or

$$X_M - \delta_{min} \leqslant X_T \leqslant X_M + \delta_{max}.$$

In the case of symmetrical tolerance, $\delta_{max} = \delta_{min} = \delta$, one writes

$$X_T = X_M \pm \delta.$$

For example, $51.3^{+0.2}_{-0.4}$ means that the true value 51.3 lies between 51.5 and 50.9. An experimental value measured to an accuracy of 5 may be denoted as

$$X_T = X_M(1 \pm 0.05).$$

DECIMAL ACCURACY

In the experimental world the quantity (for example) 51.3 mm never signifies the ideal value 51.300000 ... mm to an infinite number of significant places. It means, according to the convention adopted, that the result is closer to the precise value 51.300000 ... mm than to the limiting precise values 51.200 000 ... mm or 51.4000 mm. Similarly, the measurement 51.30 mm should be understood to be closer to the precise value 51.300 00 ... mm than to the precise values 51.290 000 ... mm and 51.310 000 ... mm. In the former case the measurement was taken with an accuracy of $\frac{1}{10}$ mm (ordinary vernier scale) and was found to lie between 51.25 and 51.35 mm, whereas in the latter case the same measurement was accurate up to $\frac{1}{100}$ mm (micrometer) and the reading lay between 51.295 and 51.305 mm.

ROUNDING OFF

An observer often wishes to round off his experimental results or to approximate his calculations on account of the uncertainty of the last figures of the result. One tends to keep two uncertain figures in any intermediate reading, observation or result, but to restrict the final result to only one doubtful figure. As an example, imagine a U-tube mercury manometer equipped with a vernier scale capable of providing readings with an accuracy of $\frac{1}{10}$ mm. A routine reading was taken and recorded as 1235.4 mm Hg. However, it was noticed that some low frequency pressure fluctuations of the order of ± 1 mm Hg were present, making the last two figures of the above result uncertain. The question now arises how should one round off the reading in order to provide reliable information?

There are a number of rounding-off rules whose primary purpose is to reduce the cumulative error. The one that is recommended below is probably most common.

Increase the last retained digit by one if the adjacent digit to be dropped is:

(1) > 5;
(2) $= 5$ and is followed by digits > 0;
(3) $= 5$ and is not followed by significant digits, or followed by zeros only, and preceding digit when increased by one becomes even.

EXAMPLES

$13.36|712 \to 13.37$; $13.36|513 \to 13.37$; $13.37|5 \quad \to 13.38$;

$13.37|500 \to 13.38$; $13.36|50 \quad \to 13.36$; $13.36|412 \to 13.36$.

ACCURACY AND PRECISION

In everyday language the two words 'accuracy' and 'precision' appear to be almost synonymous. Different meanings, however, are usually given to these words in the error analysis.

The word *accuracy* is reserved to refer to systematic errors, whereas the word *precision* is related to all incidental, i.e. random, errors. A miscalibrated instrument (e.g. with a displaced scale) can be read very precisely by a meticulous observer but the results obtained could be inaccurate.†

In other words, accuracy means correctness of measurements and precision means consistency of measurements.

8.2   Classification of errors and their nature

A remark was made in the previous paragraph that errors are generally divided into two classes: (a) *Systematic* (accountable, or fixed) errors, and (b) *Random* (unaccountable, or chance) errors.

SYSTEMATIC ERRORS

The errors belonging to this class are:

(1) *Method errors:* these errors arise when a wrong or insufficient experimental method has been chosen. Measuring of one quantity in mistake for another may occur or some unrecognised effects may influence the quantity measured so that the resultant values become erroneous. Unjustified extrapolation of experimental data may also lead to method errors.

(2) *Instrument errors:* errors of this type can be caused by a faulty instrument, the misoperation of an instrument, or by using an instrument in the environment for which it was not designed. The instrument errors are frequently biased in one direction, although in some situations hysteresis effects can occur (e.g. a worn-out micrometer, or traverse mechanism).

(3) *Calibration errors:* most instruments will not yield correct results unless they are calibrated before use against a known quantity. This may involve a simple zero setting or determination of a whole calibration curve (or scale). In either case errors can creep into the calibration procedure.

† A house owner, who happened to be an engineer, decided to check functioning of his electricity meter installed by an Electricity Board. By running just one 1 kW electrical appliance (he made sure that all other appliances had been turned off) for exactly 1 hour he found, to his discomfort of course, that the meter recorded 1.37 kWh instead of exactly 1.00 kWh. The engineer repeated the experiment several times and very much the same result was obtained. His prompt complaint to the Electricity Board stated that the meter installed was *precise*, all right, but very *inaccurate* to his disadvantage. The Board sent a man to replace the meter and our engineer again applied his own test five times to find the readings were this time 1.20, 0.90, 1.00, 1.10 and 0.80 kWh. The precision of the new measurement was poor although its average reading was 1.00 kWh. Incidentally, our engineer did not complain again – he gave up!

(4) *Human errors:* human errors depend on the personal character-istics of the observer. A human may respond to a signal too early or too late; he may either overestimate or underestimate the reading. Such errors are usually fairly consistent as they are committed perpetually by the same observer at a single session. Occasional errors committed spasmodically, owing to relaxation of vigilance for example, do not apply here and are classed as mistakes.

(5) *Arithmetic errors:* arithmetical calculations involved in experi-mentation are nowadays being increasingly taken over by various automatic computing devices (computers, automatic desk calculators, slide rules, etc.). Aberrations of such devices, however infrequent, cannot be ruled out completely. Addition-ally, there may be faults in the actual calculation procedures (programs). Incorrect rounding off can also contribute to arithmetical errors.

(6) *Dynamic response errors:* it is perhaps slightly out of place to devote a separate section to the dynamic response errors. However, their significant participation in modern experimenta-tion, especially in connection with measuring time-dependent variables, warrants a separate emphasis. Unlike static response errors (static non-uniformity of action, hysteresis), dynamic response errors arise when an instrument recording a fast changing signal fails to respond linearly to the signal variation (e.g. a pitot–static tube in fluctuating fluid flows, or electrical instruments applied to non-sinusoidal electrical currents, etc.).

GENERAL REMARKS

Yet another group of errors is sometimes quoted in the literature: illegitimate errors. These are simply blunders and chaotic errors and it is best left to the reader to decide how to deal with them.

Two important categories of experimental observations should be distinguished:

(1) *Multi-sample measurements:* mean repetitive measurements of a certain quantity under varying test conditions (different observers and/or different instruments).

(2) *Single-sample measurements:* imply one reading or succession of readings executed under identical conditions but at different times.

They refer to the problem of clearing doubts about the experimental results. A repeatability test is one way of gaining confidence, but a far more reliable way is to use an entirely different method to obtain the same results or to support a conclusion.

Having defined and classified the experimental errors, as well as having commented on their physical implications, the next step is to employ numerical procedures to quantify them. As some of the procedures, although by no means difficult, may be beyond the grasp of an inexperienced reader, it has been decided to present the respective material in Appendix 2. The introductory reading of this textbook should thus remain uninterrupted by analytical discourses. On the other hand, a reader who wishes to use the present textbook as an experimental manual has the benefit of Appendix 2, where ready-to-use technical information and practical recommendations concerning the rudiments of error analysis are provided. The Appendix deals with the following items:

forms of presentation of error affected data;
evaluation of such basic parameters as: the arithmetical mean, the standard deviation, etc.;
the Gaussian law of errors;
suggestions how to reject dubious data;
testing of experimental data for deviations from the normal (Gaussian) distribution;
propagation of errors;
curve fitting.

The supporting literature provided in Chapter 8 applies directly to Appendix 2.

The material presented in Chapter 5 and Appendix 2 is meant to be for general guidance and many experimenters may be faced with problems far beyond the reach thereof and possibly beyond the reach of more advanced texts, some of which have been included in the list of literature. It is not uncommon in dealing with errors of empirical data to be faced with the situation when one's only resort is personal intuition and experience. One should not be reticent in such cases. Individual talent in treating errors backed by expertise acquired through experience is undoubtedly a valuable asset.

There is one golden rule that almost always pays off in experimentation – the rule of patience. He who experiments patiently experiments precisely.

References

[1] Beckwith, T. G. & Buck, N. L. (1961). *Mechanical Measurements*. Addison-Wesley.

[2] Beers, Y. (1962). *Introduction to the Theory of Errors.* Addison-Wesley.
[3] Braddick, H. J. J. (1963). *The Physics of Experimental Methods,* 2nd Edn. Chapman and Hall.
[4] Brinkworth, B. J. (1968). *An Introduction to Experimentation.* English Universities Press Ltd.
[5] Cook, N. H. & Rabinowicz, E. (1963). *Physical Measurement and Analysis.* Addison-Wesley.
[6] Moroney, M. J. (1963). *Fact from Figures.* Penguin Books.
[7] Pugh, E. M. & Winslow, G. H. (1961). *The Analysis of Physical Measurements.* Addison-Wesley.
[8] Richards, J. W. (1957). *Interpretation of Technical Data.* Iliffe Books.

9
Analysis and interpretation of results

9.1 The requirements

Careful planning and execution are just as important for the analysis of the experimental data as for the actual experiment. Inaccuracies can be caused by using poor techniques during the analysis. Sometimes completely wrong conclusions can be drawn from experimental readings which are themselves quite correct.

The readings taken during an experiment have to be examined and understood as they are recorded. It is no use getting a set of readings then going away to think about them. The order of magnitude to be expected for each reading should be thought about in the planning stages. Then if any reading has an unexpected value it can be investigated while the equipment is still assembled.

9.2 The need for scepticism

As discussed in detail elsewhere, an experimental reading is not to be trusted until its background has been thoroughly investigated. An instrument might lie. It might be faulty in some way or it might be unsuitable for a particular application. Its presence might interfere with the process of the experiment or it might simply be misread. Any of these sources of error might be present so a good experimenter has a doubting, questioning attitude to all the readings. However, 'moderation is the silken thread running through the pearl chain of all virtues' and obviously not all results of experiments should be mistrusted. In practice, aircraft fly and

238

power stations generate electricity more or less as planned – and their designs are based on the results of thousands of different experiments.

When a well-planned and carefully conducted experiment does give the wrong information it is possible that the readings are correct but the *wrong* experiment has been performed. For example, in an early attempt to design a retractable undercarriage for a fighter aircraft, experiments were made to assess how much the aerodynamic forces would distort the mechanism. When the prototype was designed and built using the results of these experiments, it jammed. Eventually it was found that centrifugal growth of the rapidly spinning tyre, which had been completely overlooked during the experiments, was a major problem. No amount of care in measuring the effects of aerodynamic forces could have made up for this oversight.

9.3 Preliminary work

Before the experiment starts, a way to deal with the readings must be chosen. The likely value of each measurement should be estimated and, where appropriate, a table should be prepared for recording them in an orderly manner. If a reading has an unexpected value, it should be investigated at once: either a fault has occurred or an unexpected and therefore interesting phenomenon has been discovered.

Instrument readings should be plotted as they are taken. Any regions of particular interest can then be noticed straight away and extra readings can be taken near critical points on the curve. The axes should be prepared and the expected shape sketched in before the experiment is started. This procedure will allow any serious errors to be noticed before it is too late to correct them.

By plotting the actual measured values, the need for rapid mental arithmetic will be avoided. Any corrections or conversions which need to be applied should be noted in the laboratory log book so that they are not forgotten when the information is finally analysed.

9.4 An example of 'plotting as you go'

In an experiment to measure the electrical characteristics of a domestic light bulb, the circuit shown in Fig. 9.1(a) was set up. The voltmeter read directly in volts over the desired range but the ammeter had a control marked '$X1$, $X2$, $X5$' which was set at the '$X2$' position. A quick calculation was made before the circuit was switched on showed that at its normal operating voltage of 240 V, the 100 W bulb would be expected to consume a current of 0.42 A. The current meter would be expected to have a reading of $0.42/2 = 0.21$ units. When the field was prepared for the graph a scale from 0 to 250 V was laid out for one axis and from 0 to 0.25 units for the other. (This was intended to go slightly beyond the expected meter readings at the operating voltage.) These axes are shown in Fig. 9.1(b). The point R corresponds to the rated output.

When the test was started the first three pairs of readings were as shown in Table 9.1. Plotted on the graph they gave the points A, B and C. It was immediately obvious that the curve was not heading towards point R. Investigation showed that operation of the 'X2' switch had the effect of doubling the meter deflection and not, as had been assumed, of requiring the meter reading to be doubled to obtain the current. Many meters have ambiguous markings; they should always be used with caution. The rated output should therefore have been represented by a point R_1 at 240 V, 0.833 units on the current meter which lay in the track of ABC. Caution is always required when using such range switching controls.

What if the readings had just been taken without thought and plotted later? The current for each voltage used would have been taken as twice the meter reading, i.e. four times the correct value. In this simple case the error

Fig. 9.1. (a) Circuit used for measuring the electrical characteristics of a light bulb. (b) Graphical monitoring of measurements taken during the light bulb test. The point R represents the normal operating condition. It is obvious that a mistake has been made and the measured currents are much higher than expected.

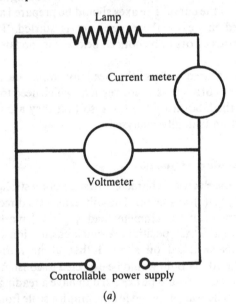

(a)

would probably have been spotted without difficulty. In a more complicated situation where the readings are used in lengthy calculations, such errors might pass unnoticed or at least cause worry and extra work.

The graph in Fig. 9.1(*b*) was made in the laboratory log book; it was not the one which finally appeared in the report on the experiment as no corrections had been made at that stage for instrument calibration or for the effect of the potential drop across the ammeter.

What about the cases where a reliable point on the curve cannot be obtained so readily? The manual should of course be studied before a strange piece of equipment is used. Then the effect of all instrument range switches, which might be marked ambiguously, should be checked

Table 9.1. *Readings taken during the early stages of the light bulb experiment*

	A	B	C
Voltmeter reading	20	40	60
Ammeter reading	0.12	0.22	0.31

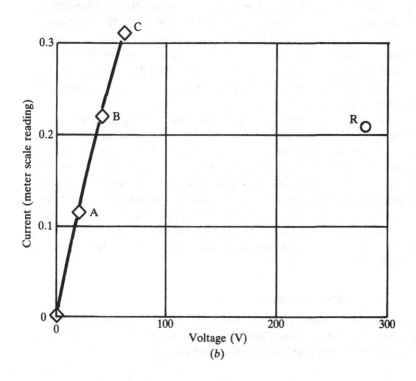

(*b*)

experimentally. This can be done by applying a constant input and watching how the reading changes when the switches are operated. If possible, known values should be fed to each of the instruments to check that the readings are of the right order. This is not to calibrate the instruments but to detect mistakes in understanding their scales.

An attempt should always be made to obtain an estimate of reasonable upper and lower limits of every quantity which is to be measured. For example, in a boiler test most but not all of the heat available in the fuel should go into the steam so any value outside the range of, say, 50–95% would be suspect. This sort of accuracy is sufficient to detect gross errors and it can be coupled with an idea of the trend to be expected when measurements are made over a range of conditions. In the boiler the smallest proportion of the heat would be wasted when operating near the design output level; much more heat would be wasted at very low and at very high steam production rates.

9.5   Pilot experiments

An experiment should be preceded by a preliminary trial of the apparatus. The main purpose of this initial test is to check that the equipment is working correctly and to give the experimenter practice in adjusting the controls and reading the instruments. At the same time the opportunity should be taken to collect and process a set of readings. It can then be seen that the measurements are giving sensible and sufficient information. Provided they lead to satisfactory conclusions, the pilot readings can then be used to provide a check on the readings taken and plotted during the main experiment.

To save time and to safeguard apparatus from damage during teaching experiments, the pilot tests are normally performed in advance by the instructor. The students are then advised about instrument ranges and settings, any particular phenomenon to be observed, and the sort of readings to expect. The students can then quickly gain experience of handling the apparatus and of taking readings. However, they ought to take the trouble to think about the pilot tests and discuss them with the demonstrator. Otherwise they will obtain little value from the experiment and will only perform tasks which could be delegated to technicians or automatic data loggers.

9.6   The law of averages

Frequently, when readings are inconsistent, an average value of a set of measurements is calculated. This practice is not always sensible; the use of the mind should precede the use of the mean.

Returning to the light-bulb experiment, the test was resumed and the readings plotted in Fig. 9.2. The curve passed close to the point R', predicted for the rated value, and up to the point corresponding to the highest voltage applied during the experiment. Corresponding current and voltage readings were then taken as the voltage was reduced back to zero. These were found to lie on the curve through C' which is shown in the diagram. A procedure had to be found to obtain the true characteristic curve. There were two distinct sets of values, one from the increasing and one from the decreasing voltage. It would have been simple to average the two currents for each voltage but this would hide the differences and lead the experimenter away from some interesting and perhaps important features of the bulb's behaviour. So, instead, the various possible explanations were listed as shown in Table 9.2.

All the possibilities (a)–(d) which could be thought of were listed together with their consequences. This table allowed further experiments to be planned to find the cause of the discrepancy between the two curves. It was eventually found that a combination of (b) and (c) had occurred. Both the temperature of the wiring and ageing of the bulb had affected the current. No original thinking was required to investigate this problem, just a logical approach and a desire to understand what had happened.

Fig. 9.2. Readings taken during the light bulb test. The values are now of the right order but there is a lower current for each voltage when the voltage is being decreased. It is wrong to assume that the correct value of current lies half-way between the two curves.

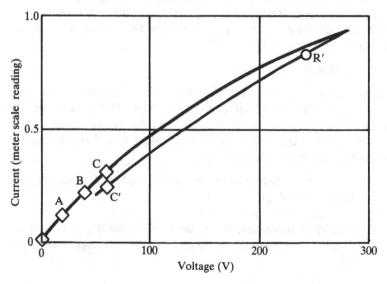

Table 9.2. *Possible explanations for the light bulb current readings*

	Possibility	Consequence
(a)	The current/voltage curve has a hysteresis loop	Raising the voltage again would cause current values lying on the line *OC*. The circuit must have unique electrical properties
(b)	The bulb and wiring were hotter at point *C′* than at point *C*	Repeating the test but passing quickly over the region *CR′C* would reduce the difference
(c)	The bulb was damaged by overloading	Readings in a new test would lie along the line *OC′*. Holding at the maximum voltage would cause a further change
(d)	A fault has developed in one of the instruments	The test would not be repeatable
(e)	...?...	...?...

Once the error caused by heating of the wiring had been discovered, its magnitude could be found and a correction made. Further corrections were necessary for the calibration errors in the two instruments and for the voltage drop across the ammeter.

Another type of situation where a simple average would have been wrong arose during the measurement of the natural frequency of a slowly vibrating structure. The time for 50 complete vibrations was measured on a digital stopwatch. The measurement was made six times resulting in the following values. (The actual values were measured to the nearest one hundredth of a second but, for clarity in explaining the principles, they are shown rounded to the nearest tenth of a second.)

Time = 58.9, 60.1, 58.7, 62.6, 59.5, 60.2 s.

What was the average time? The obvious answer would be the total time of 360 s divided by the number of readings, i.e. 360/6 = 60.0 s. However, the fourth value (62.6 s) seems out of line with the other figures. It should be considered carefully before it is included in the calculation. Possible reasons which might explain the odd value are:

(a)   human error such as miscounting the number of cycles or being slow to switch off the clock;

(b)   an isolated fault in the clock;

(c)   a different frequency; the mode of vibration might have been different or a significantly different amplitude might have been used.

The best course of action in such a case is to take a few more readings and then apply the James Bond dictum: 'Once is happenstance, twice is coincidence, three times is enemy action'. The worst way of handling scattered data is to take the average without carefully considering why the values are different. An average is wrong if it includes incorrect values. An average is meaningless if the quantities included represent different situations: for example, there is no sense in the statement that the average creature in a certain zoo has one wing and three legs.

When no further readings can be taken, the decision whether or not to discard a particular reading can be made by using the method described in Appendix 3. Assuming normal distribution, the relevant figures for the present example are:

$$\text{standard deviation} = 1.41 \text{ s};$$
$$\text{deviation with } 50\% \text{ probability} = 2.4 \text{ s};$$
$$\text{minimum probable time} = 57.6 \text{ s};$$
$$\text{maximum probable time} = 62.4 \text{ s}.$$

It follows that the fourth value of the time should be rejected because its deviation is too high. However, it must be stressed that such statistical calculations have to be used with caution. They cannot give correct values unless the correct form of the distribution curve is known. Unthinking calculation of averages is all too easy when a large number of readings are being collected automatically on a data-logger. In such cases it is most important for the experimenter to supply the necessary judgement to complement the automatic equipment.

An engineering approach to dealing with scattered results such as these is to decide in advance that the highest and lowest readings will be discarded. If the discarded readings are wrong then the average will be improved. If they are correct then the average will not be changed very much so no harm will be done. When a large number of readings are being averaged, then one-tenth of their number can be discarded at each end of the range.

9.7 Tables of possibilities

The experimenter investigating the light bulb's behaviour very sensibly made a table of possible explanations of the results. Such a table, normally made in the laboratory log book, ensures a logical investigation. It can also be invaluable when further work is done on a particular problem after a gap of some months. Even in the simple example described, it can be seen how a table enables different possibilities to be isolated then eliminated or confirmed separately.

It is all too easy to blame instrument error for any strange results; however, it is found in practice that intermittent errors are quite rare so it was put low down on the list of possibilities. This encouraged thorough investigation of the alternatives. Small calibration errors are common, large errors are normal, particularly when an instrument is used near the extremes of its range. What is unusual is exactly the right isolated fault to explain such strange experimental readings.

When further tests are made to investigate the various possibilities, the table should be worked through in order. Frequently, more than one of the phenomena are present at the same time. They might either reinforce or tend to cancel each other. A necessary line in the table is (e) – a blank space waiting for any further thoughts which might occur.

9.8   Faults in measurements

Before using experimental readings, it is necessary to decide:

   (a)   how much the instrumentation interferes with the process being studied;
   (b)   whether the experimental arrangement models the real situation;
   (c)   what errors the instruments have (these are discussed in Chapter 5 and Appendix 2).

It is accepted by thermodynamicists and other philosophers that, whenever a measurement is made, the value being measured is altered. In many cases the alteration is completely negligible; for example, the act of measuring the temperature of the air in a room will make such a small difference to its value that the resulting error will be much less than the discrimination of the most sensitive thermometer. However, sometimes the error can be significant; the introduction of a mercury in glass thermometer would cause quite a noticeable drop in the temperature of a cup of coffee. Sometimes the effect is enormous. Once a large aircraft crashed because overheating of a vital oil supply led to a total electrical failure. The cause of the overheating was a set of temperature measuring probes which obstructed the oil circuit.

On a less disastrous level, it is always difficult to measure the rapidly changing pressure in an internal combustion engine cylinder. The pressure transducer alters the volume, shape and heat transfer characteristics of the combustion chamber so that the combustion process in the instrumented engine is bound to be different from normal.

Since the magnitude of the effect cannot be calculated, it must be measured. One way of doing so involves repeating the measurements with a second transducer added. The difference between the two sets of readings allows the effect of the addition to be estimated. This method of finding how much difference the instrumentation makes can be applied in a large number of different situations. For example, provided sufficient identical cups of coffee were available, the correct temperature could be found by reading the apparent temperatures of different cups with one, two, three and perhaps four thermometers inserted. A graph of apparent temperature against number of thermometers could then be plotted and extrapolated back to zero. Many extra readings are required to make corrections in this way. The number can sometimes be reduced if just a few, carefully chosen, experiments are made with the extra instruments to produce general information about the errors.

A similar technique involves varying the value of one of the experimental parameters and extrapolating to a perfect case where errors are zero. In this way the unwanted effects of heat losses, electrical resistance, friction and distortion can often be overcome. One example shows the general method. The 'heat transfer coefficient', h, between a solid and a fluid is usually measured by passing the fluid through an electrically heated pipe. The internal surface area of the pipe is known; the temperatures of the fluid and the internal surface of the pipe can be measured. The rate of heat flow is equal to the electrical power supplied to the heating coils. Hence h, which is the heat flow per unit area per unit temperature difference between the solid and the fluid, can be calculated.

An error is inevitable because some of the energy from the heating coil is lost through the surrounding insulation or by conduction through the ends of the pipe. The amount of heat lost in this way will depend on the power at which the circuit is operating. Measurements can be made at different power settings and the apparent value of h calculated in each case. Fig. 9.3(*a*) shows how h appeared to vary for one particular arrangement. At zero power there would have been no heat losses. Hence by extrapolating the line through the measured points back to zero power the true value of h can be found. Some advice on extrapolation is given later, in Section 9.14.

Fig. 9.3(*a*) was plotted with an expanded vertical scale and a false origin. This allowed precise plotting and extrapolation to be achieved. However, the impression given by such a graph is quite misleading. The relative importance of the correction can be seen more clearly in Fig. 9.3(*b*) which is plotted to a natural scale. In science and engineering (as distinct from advertising) any graph with a false origin should be accompanied by a warning note to avoid misleading the reader.

Many experiments are quite different from the situations they are supposed to simulate. If the differences are realised an allowance can be made for their effect on the measurements. Occasionally they are overlooked completely and totally wrong conclusions are drawn about the phenomenon being investigated.

The likelihood of a certain pipe union being damaged by accidental overtightening was once investigated by measuring the torque–angle of rotation relationship in a torsion testing machine. When assembling the union, which is shown in Fig. 9.4(a), the nut N is rotated so that it draws the two elements A and B together in order to make a leakproof joint. If an

Fig. 9.3. (a) Graphical correction for heat losses. The losses in this case are a function of the operating power. By varying the power, different errors lead to different incorrect values of the heat transfer coefficient. The correct, error-free, value is obtained by extrapolating back to zero power. Note the misleading effect of the false origin. (b) The effect of operating power on apparent heat transfer coefficient. When the data shown in (a) is replotted using a natural scale, a more accurate impression of the apparent variation in h is obtained. This graph gives a good qualitative picture but does not allow precise extrapolation.

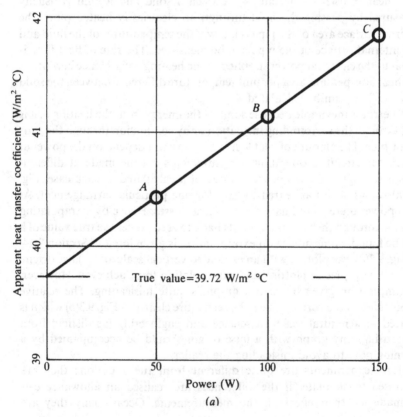

(a)

excessive torque is applied, the union fails when the nut fractures at the point *F* shown in Fig. 9.4(*b*). During the tests, the component *B* was gripped in one side of the torsion testing machine. A box spanner was fitted over the nut and driven by the other side of the machine. The measured torque was far higher than it should have been because the nut rubbed on the jaws of the testing machine as shown in Fig. 9.4(*c*). This error was not noticed until after time and material had been wasted obtaining many completely unreliable results.

9.9 Tabular analysis

When a large amount of experimental data has to be understood, planning is necessary to avoid mistakes and to save time. Frequently the data and calculations are repetitive, as when measurements are made of the response of a system to a range of input variables. A tabular layout is then useful. The

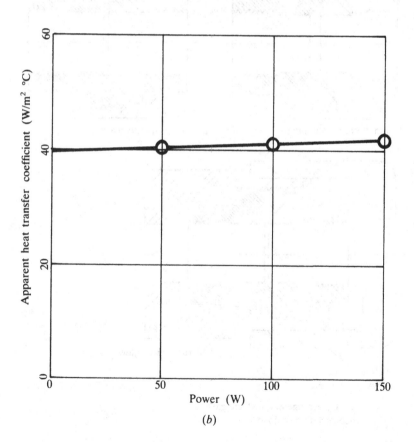

(*b*)

calculations can be performed manually for relatively simple cases but often it will be worth using a computer. A special program might be necessary but time and effort can be saved by using one of the many 'spreadsheets' which are available. Purchasers of these general-purpose tabular programs should make sure that they choose one with a complete range of mathematical functions. Some of the cheaper spreadsheet programs are intended for simple book-keeping calculations and can only perform simple arithmetic.

Fig. 9.4. Testing of a pipe union: (a) union before tightening; (b) failure due to overtightening; (c) testing arrangement which gave incorrect failure load.

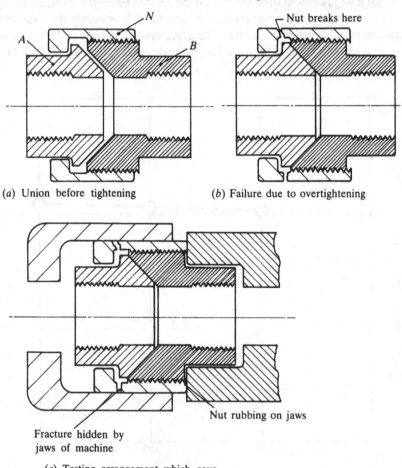

(a) Union before tightening   (b) Failure due to overtightening

(c) Testing arrangement which gave
     incorrect failure load

Where a spreadsheet or other computer program is to be used to process the experimental readings, it should be set up in advance. The readings can then be keyed directly into the computer so that processed results are immediately available. These results can be used instead of the raw readings for plotting a graph during the experiment. In this way the one graph can be used to monitor several different instrument readings by monitoring their combined effect.

Whatever method is used to construct a table of processed results, at least one set of figures should be calculated individually as a check that the correct operations are being performed.

An example of the tabular method of calculation is shown in Table 9.3. The figures are from a test to find the power output and fuel consumption of a small petrol engine operating at a fixed throttle setting over a range of speed. The first four lines are the measurements and the remainder are derived results. Intermediate stages in the calculation are recorded even when they are not needed immediately. For example, column *H* shows the fuel consumption per hour in case it might be useful later and to help in checking for mistakes. Only a few significant figures are given in the table but accuracy is not lost as the full figure is carried in the calculator or computer for use in the later stages of the calculation.

By glancing down the columns of figures, there is a good chance of spotting mistakes as the values will normally change smoothly. For example, the effect of a mistake in recording the brake deadload can be seen in Table 9.4. Because the results are recorded in order, the calculated values should change fairly smoothly. Examination of column *G*, for example, shows that there is an unexpected value of power in row 13. This was caused by an incorrect entry in column *B*. Where the calculations are made by hand, arithmetical errors might occur; these can be detected in the same way.

The spreadsheet used for this example has a bar chart facility. This can be used to show the general form of the results as the experiment progresses. The main purpose of the graph is to show points of interest. For example, in the present case it shows that the range of speeds have been sufficient for the maximum power point. The graph can also show mistakes such as the mis-keying of the brake deadload in slot *B*13.

9.10 Accuracy of calculations

Once any numerical data have been manipulated on a calculator or computer, it will be necessary to round off the results to a suitable number of digits. Leaving the full eight or ten figures in the results is misleading and,

Table 9.3. *The results of a test on a small petrol engine recorded and analysed by a spreadsheet program*

	A	B	C	D	E	F	G	H	I
1	Speed	Brake dead wt	Brake balance	Time to use 50 g	Brake net	Torque	Power	Fuel consump.	sfc
2	rev/min	kg	kg	sec	kg	Nm	kW	g/hr	g/kWh
3									
4									
5	Data	Data	Data	Data	B-C	$4.905 \times E$	$AF/9550$	$180E3/D$	H/G
6									
7	800	75	5.8	37.2	69.2	339.4	28.4	4839	170.16
8	900	75	4.7	34.2	70.3	344.8	32.5	5263	161.95
9	1000	75	5.6	32.5	69.4	340.4	35.6	5538	155.37
10	1100	75	7.1	28.9	67.9	333.0	38.4	6228	162.35
11	1200	75	10.2	28.4	64.8	317.8	39.9	6338	158.68
12	1300	65	3.8	25.8	61.2	300.2	40.9	6977	170.72
13	1400	65	7.9	24.1	57.1	280.1	41.1	7469	181.90
14	1500	65	12.7	22.9	52.3	256.5	40.3	7860	195.06
15	1600	50	2.5	18.7	47.5	233.0	39.0	9626	246.58

Note: The first four columns are the experimental readings. The remainder were calculated automatically as the data were keyed in by means of the formulae shown in row 5. By preparing the spreadsheet in advance, it was possible to obtain the derived results immediately the readings were taken. The bar chart at the right-hand side shows the power output. A false zero has been used to allow small differences in power to be seen. If extra readings are taken, then extra rows of figures can be inserted as required. This will make it easy to spot irregularities but any time dependent phenomena might be lost.

Table 9.4. *The effects of mistakes and errors on the engine results*

	.....A	.....B	C	.....D	.....E	.....F	.....G	.....H	I	
.....1	Speed	Brake dead wt	Brake balance	Time to use 50 g	Brake net	Torque	Power	Fuel consump.	sfc	
.....2										
.....3	rev/min	kg	kg	sec	kg	Nm	kW	g/hr	g/kWh	
.....4										
.....5	Data	Data	Data	Data	B–C	$4.905 \times E$	AF/9550	180E3/D	H/G	
.....6										
.....7	800	75	5.8	37.2	69.2	339.4	28.4	4839	170.16	******
.....8	900	75	4.7	34.2	70.3	344.8	32.5	5263	161.95	**************
.....9	1000	75	5.6	32.5	69.4	340.3	35.6	5538	155.37	*********************
.....10	1100	75	7.1	28.9	67.9	333.0	38.4	6228	162.35	***************************
.....11	1200	[74.8]	10.2	28.4	64.6	316.9	39.8	6338	[159.17]	*****************************
.....12	1300	65	3.8	25.8	61.2	300.2	40.9	6977	170.72	******************************
.....13	1400	[75]	7.9	24.1	67.1	329.1	48.3	7469	154.79	**
.....14	1500	65	12.7	22.9	52.3	256.5	40.3	7860	195.06	****************************
.....15	1600	50	2.5	18.7	47.5	233.0	39.0	9626	246.58	**************************

Note: A mistake in keying in the experimental data will be noticed when the calculated results are inspected. In line 13 the wrong value of brake deadload has been used. The resulting discontinuities in columns *E*, *G* and *I* can be seen. The bar graph makes such large errors easily noticeable.

The spreadsheet can be used to find the sensitivity of the calculated data to small errors in the measurements. An error of 0.2 kg has been deliberately introduced into slot *B*11. By comparing the specific fuel consumption (slot 11) with the corresponding value in Table 9.3, the seriousness of such an inaccuracy can be assessed without the need for complicated analysis. Incorrect values are bracketed and referred to in the text.

therefore, wrong. The full capacity of the calculator is normally used for all the calculations; the rounding off is only performed on the final results. This practice saves the accumulation of rounding errors in successive stages of the calculation.

The final degree of accuracy will depend on the accuracy of the experimental readings and on the number and type of the calculations performed on them. Sometimes the resulting accuracy can be calculated, sometimes judgement is necessary together with numerical experiments. An example will show the type of approach to use.

Suppose that the frequency is required for the low-speed vibration discussed earlier. Ignoring the fourth reading, the average time is found to be $297.4/5 = 59.48$ s. The frequency is, therefore, $50/59.48 = 0.840\,618\,695$ Hz. An estimate of how many of the decimal figures are significant can be made by considering what would have happened if by chance only four readings had been taken. Missing out the highest and lowest in turn leads to possible values of $0.843\,170\,32$ and $0.837\,871\,805$ Hz respectively. It is obvious that the third decimal place is quite unreliable in this case and the frequency is 0.84 Hz.

Similar numerical experiments can be made on the petrol engine results table. For example, the weights used for the brake dead load might only be accurate to 0.2 kg. It is a simple matter to see how much an error of that size would affect the calculated data in each column. Normally such an investigation would be made by feeding in data by hand to represent the limits of possible error. This has been done in row 11 where the resulting difference in specific fuel consumption can be found by comparing slot $I11$ with the corresponding slot in Table 9.3. Before any such numerical experiments are made, the original data must be saved and labelled. It is very easy to try several alterations to the input table and then to forget the original values.

Sometimes it is worth including tests in the computer program which automatically demonstrate the sensitivity of the calculated results to errors in input values. The complication might be worthwhile when a number of different readings are being combined in such a way that it is not immediately apparent how a variation in a single reading will feed through to the final result.

Calculations should be carefully preserved for future use. They should be made in a laboratory log book or in a special calculation book. In either case they should have sufficient explanatory notes. Where calculations are done by computer, extra care is needed to keep good documentation. If a program is likely to be modified later, a copy of the original version should be saved. It should be possible to understand any calculations several years later in case there is any query to answer.

9.11 Graphical analysis

Frequently, experimental data can be analysed graphically with a saving in labour and time compared with numerical analysis. An incidental advantage of graphical methods is that they give a clearer picture of trends in behaviour and of scatter in experimental results. Indeed, when organising a computer program to handle the data from a long series of experiments, it is sensible to analyse the first few sets of results graphically to make sure that no special features of the results disappear for ever inside the computer.

Points can be plotted on a graph with the same precision as most instruments give. If a fine line is drawn as a good fit to the points it can average out scatter and so improve on the accuracy of individual readings. Any points which do not fit into the general pattern can be spotted. After investigation they can be studied further or ignored as appropriate.

The potential accuracy of graphical analysis can be destroyed by three common mistakes. All three mistakes were made when Fig. 9.5(*a*) was plotted. This graph shows readings of force and extension which were recorded during a tensile test on a sample of an aluminium alloy. In such a test, the load is plotted against the extension as the test piece is slowly stretched until it fractures. Two of the figures required from the graph are the slope of the initial, straight, part of the curve and the load at which non-linear behaviour starts.

The most obvious mistake was to measure the slope of the graph as it was originally plotted. For the slope to be measured accurately, scales should be chosen to ensure that the line is at about 45° to the axes. In many cases, including this one, only part of the range can be included in the graph. If the slope is required at several positions, then several sections of the graph will have to be drawn separately.

When the initial part of the graph was redrawn using a more suitable scale as shown in Fig. 9.5(*b*), the second mistake was shown clearly. When the graph was drawn it seemed reasonable that, as zero deflection required zero force, the origin was the most accurate point on the graph. Therefore the graph was drawn from the origin through the plotted points with about the same number on each side. Unfortunately the common, and natural, assumption about the importance of the origin was wrong. At the best the origin is just another measured point. Frequently, as in this case, it is the least reliable point of the lot. Many instruments have errors caused by friction or backlash in the mechanism or sometimes by inaccurate zero adjustment. When any of these errors is present, all the readings will be in error by a fixed amount. The best straight line through the data, but with the origin ignored, is shown in Fig. 9.5(*c*).

Since the nature of the problem requires that the line must pass through the origin, it is necessary to move the reference grid of the graph until the origin lies on the line. Whether the origin should be O_1 or O_2 or at some other point cannot be decided reliably without further experiment. In this case no more information is available so it is necessary to make the most reasonable choice. If O_1 were to be chosen, it would imply that the deflection measurements were correct but that the experiment started with a negative force applied to the test piece. This does not seem a very likely possibility. The choice of O_2 would mean that no deflection reading

Fig. 9.5. (*a*) Tensile test readings as originally plotted. The horizontal scale is too small to allow the slope of the straight line to be measured accurately. (*b*) The linear part of the tensile test results drawn with a more suitable horizontal scale. The poor fit is caused by drawing the line through the origin. (*c*) A better fit to the linear part of the tensile test data. The origin has been ignored when drawing the line. It can be seen that there is a zero error in one or both of the instruments. When measuring the slope, using a large length of the line such as $A'B'$ gives a more accurate result than using a small section such as AB. (*d*) An enlarged view of the top of the linear part of the tensile test graph. It is more unlikely that the change in material behaviour would occur exactly at a measurement point. (*e*) A more accurate version of the top of the linear part of the tensile test graph. The two parts of the graph were drawn independently and extrapolated to the point of intersection.

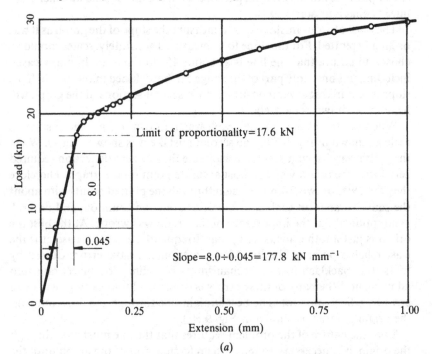

Limit of proportionality=17.6 kN

Slope=8.0÷0.045=177.8 kN mm^{-1}

(*a*)

occurred until a certain amount of clearance was taken up and all the deflection readings should be increased by O_2O above the measured values. It is possible, however, that the true origin lies at an indefinable point such as O_3. Zero errors are just as likely on non-linear graphs. To avoid

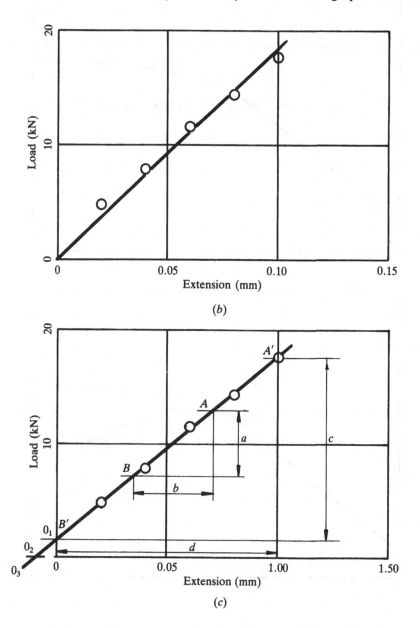

(b)

(c)

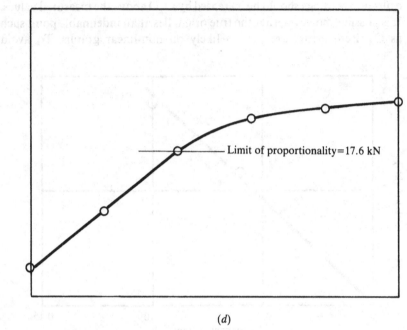

Limit of proportionality=17.6 kN

(*d*)

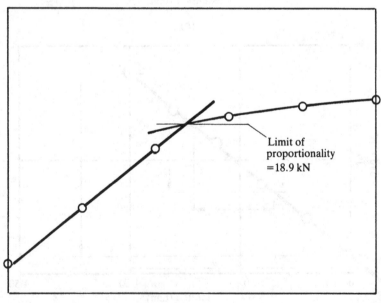

Limit of
proportionality
=18.9 kN

(*e*)

distorting the curve the origin should be ignored (or at least not given special emphasis) when a graph is drawn.

The final mistake occurred when estimating the load at which the graph became non-linear. This region of the graph is shown enlarged in Fig. 9.5(*d*). The mistake was to assume that a point of interest would be exactly at a point of measurement. In fact the two are most unlikely to coincide. The same mistake is often seen in drawing maxima and minima or points of inflection. A much better drawing of this region of the graph is shown in Fig. 9.5(*e*).

The exact details of the start of the non-linear part of the curve cannot be found from the limited number of plotted points. A pilot experiment should have been used to find the approximate position of this region of interest, then it would have been possible to take readings at shorter extension intervals in this region.

To obtain the modulus of elasticity of the material, it was necessary to measure the slope of the graph. Even after the relevant part of the curve had been drawn correctly (in Fig. 9.5(*c*)), it was still possible to lose accuracy. Mathematically it would seem that by choosing any two points such as A and B and calculating a/b the slope would be obtained. However, when reading off the coordinates of A and B there will be a certain unavoidable error. The magnitude of this error can be reduced by spacing A and B as far apart as reasonable at, for example, A' and B'. The error in either a or b can be reduced by making it a convenient whole number. By choosing the horizontal distance the division can also be simplified, frequently with the avoidance of any rounding-off error. It follows that the value of c/d gives a more accurate figure for the slope than a/b.

When measuring intervals such as a, b, c and d, the scale of the graph should be used rather than a separate rule. This is because the printing process and paper shrinkage can cause variations in the size of the markings. Furthermore, these effects are directional; the squares on a sheet of graph paper might not be exactly square in practice. When extreme accuracy of plotting is required, squared paper is not used. Instead a steel rule is used for measurements and the graph is drawn on a metal or stable plastic sheet. Measurements on 'log–log' graphs are an exception to the above advice, as will be seen later.

9.12 Fitting curves to experimental data

It is not easy to draw the 'best' curve through a series of points such as those shown in Fig. 9.6(*a*). A choice has to be made between four possible assessments of the situation. This choice must be a conscious one and the correct assessment will depend on the source of the data as well as the positions of the plotted points.

A simple curve can be drawn by eye or with the help of numerical curve fitting techniques. For example, the best fitting parabola is drawn in Fig.

Fig. 9.6. (*a*) Experimental data which appears to have a lot of scatter. There is no single correct curve which can be drawn through such a set of points. Several choices are possible depending on the circumstances. (*b*) The best fit parabola through the scattered data. If this is the correct curve, two of the points are very inaccurate. (*c*) The best fit cubic through the scattered data. This curve is more complicated but is still not a good fit. The shape of the graph can hardly be justified by the limited available data. (*d*) The effect of removing one of the points. A simple curve such as this best fit parabola can now be drawn to give a good fit to remaining points. If it is likely that the missing point contained a mistake or a large error, then this is the correct curve. (*e*) A discontinuous line through the scattered data. Such a graph might be correct if a change in zero error occurred part way through the test or if the process being measured was discontinuous.

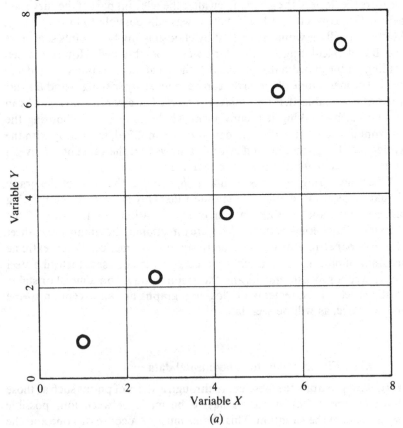

(*a*)

9.6(*b*). This particular curve can only be correct if the measured values of *X* and *Y* are liable to have the large errors which can now be seen.

If the fit of the curve shown in Fig. 9.6(*b*) is not good enough, then the obvious step is to draw a more complicated curve. The cubic in Fig. 9.6(*c*) involves smaller errors but can only be correct if the relationship between *X* and *Y* just happens to have a point of inflection in the middle of the plotted range and maximum and minimum values just outside the ends of the range. By drawing a polynomial or Fourier curve of sufficiently high order, the plotted points could be fitted exactly. It is most unlikely that such a curve would be the correct one, however. The test to apply before choosing a complicated curve is to remove one or more points. Since the original choice of values *X* will have been arbitrary, the omission of one point should not make any real difference. If the proposed shape now looks unnecessarily complicated it is probably wrong. For example, if no reading

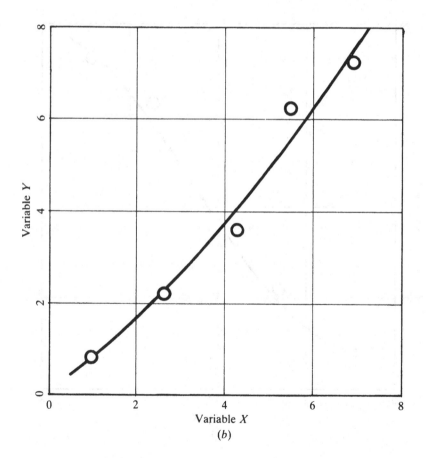

(*b*)

had been available for $X = 5.50$ (the fourth point) then the upward bulge in the right-hand part of the curve would have no purpose. The complexity of the curve in Fig. 9.6(c) is only required for this one point. This is not to say such a curve cannot be correct. There is not sufficient evidence to make a decision without further points in the regions of $X = 5$ and $X = 6$.

The next possibility to consider is that, rather than most of the points having large errors, just one point contains a serious mistake. If, for instance, the fourth point is discarded, a simple parabolic curve can be drawn (Fig. 9.6(d)) to give a good fit with the remaining points.

The final possible explanation for the 'scatter' is that there is a discontinuity in the curve. An instrument fault could have started between the third and fourth readings. This could have been caused, for example, by accidental interference with a zeroing device or by changing the range setting on an incorrectly calibrated instrument. A discontinuous curve can

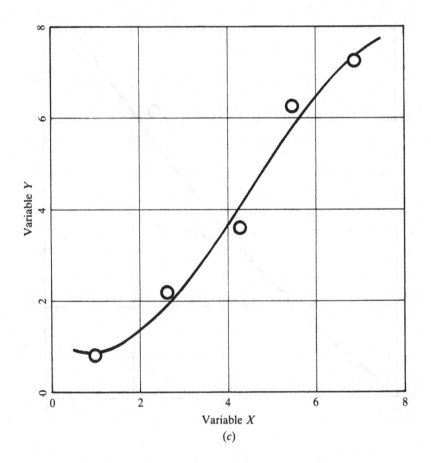

Variable *X*

(*c*)

represent a real discontinuity in some physical systems. For example, a vapour might exist in an unstable manner at a temperature below its normal condensation point, then suddenly condense with a consequent sudden volume change. If one of these circumstances were to apply to the present case, the correct curve might be similar to that in Fig. 9.6(*e*) – a *discontinuous straight line.*

Points cannot be lightly discarded whenever they fail to fit preconceived notions of the shape of the graph or just because they make the graph look untidy. They must first be carefully checked and the source of the mistake or the unusually large error found. This is made easier if the recommended practice of plotting results before finishing the experiment is followed. Drawing a curve through scattered data is in many ways similar to the process of finding the average of a number of readings which was discussed earlier. Values which show large deviations from the rest should be checked

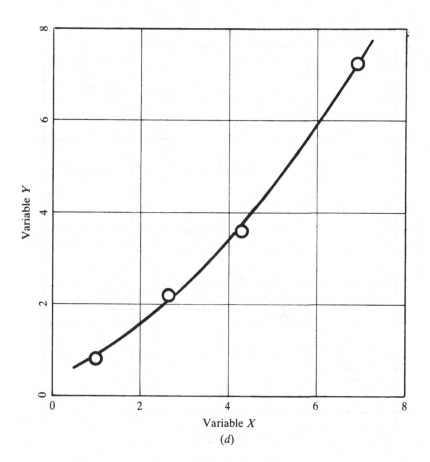

(*d*)

carefully and possibly discarded. They should not be allowed to distort the pattern of results.

The above uncertainties about the shape of a curve can sometimes be overcome by taking a large number of experimental readings but this is expensive and time consuming. A better approach is to plot the data to scales chosen so that a straight line is formed. As explained in Chapter 2, some trial and error work is required but an intelligent approach can reduce this to the minimum. Advantage should be taken of any possibility of qualitative reasoning about extreme values, particularly when only a small quantity of experimental data is available.

For example, a number of experiments were made to find out how the different dimensions and the choice of materials affected the deflection of a certain type of structure under the action of a certain load. Because only two materials were available only two results could be obtained when the

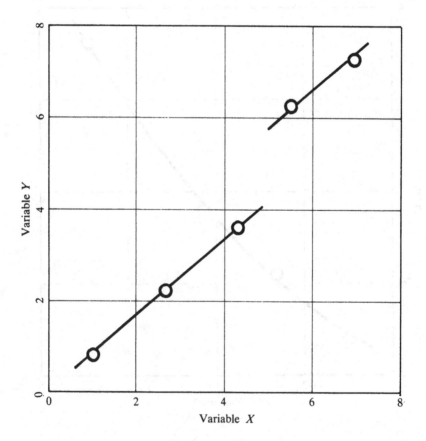

Fig. 9.7. (*a*) A graph drawn through limited data. Only two points, *A* and *B*, are available and it might be thought that the only possible curve is the straight line shown. However, this leads to absurdities at points *C* and *D*. (*b*) The same data plotted with a different horizontal axis. This straight line leads to logical values all the way along its length. It is easy to measure the slope of such a line and hence find a mathematical relation linking the variables. (*c*) The correct shape of curve through the two original points. This was drawn with the aid of the expression found using the previous graph.

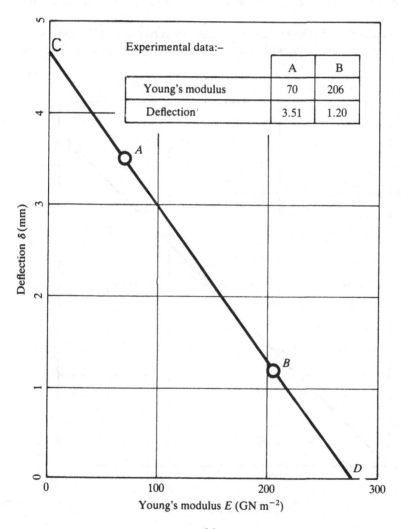

Experimental data:-

	A	B
Young's modulus	70	206
Deflection	3.51	1.20

Deflection δ (mm)

Young's modulus *E* (GN m⁻²)

(*a*)

effect of Young's modulus was being investigated. These were plotted as shown in Fig. 9.7(a). It might be thought that the only graph which could be drawn would be the straight line CD, as shown. However, this would have implied that, for a certain modulus D, the deflection would always be zero and for a greater modulus the deflection would actually be negative. At the other extreme, the use of a material with no resistance to deformation, that is, with a Young's modulus of zero, would cause the structure to have a finite deflection of $OC = 4.7$ mm. It was clear that the straight line, or indeed any curve cutting the axes, would not satisfy the physical realities of the structure.

It was decided that some sort of hyperbola was required. The results were plotted again, this time with $1/E$ as the independent variable. The two

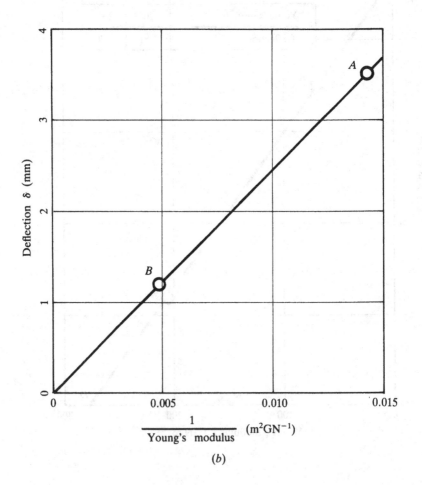

(b)

measurements were found to lie on a straight line through the origin as shown in Fig. 9.7(*b*). The origin corresponds to an infinite value of Young's modulus, i.e. to a completely rigid material, and to zero deflection, which is a logical combination although physically unattainable. Therefore it was decided that the straight line was correct. The conclusion was drawn that the deflection was inversely proportional to Young's modulus, a result which could, in fact, have been ascertained from dimensional analysis in

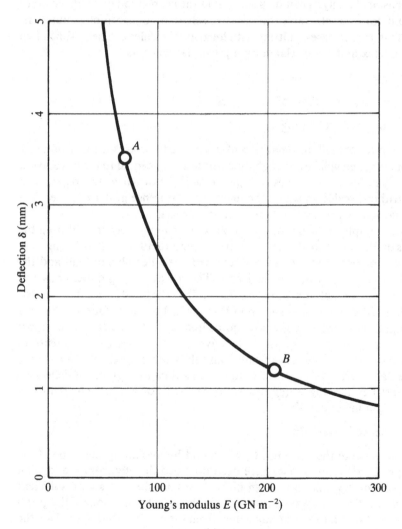

(*c*)

this case. By measuring the slope of the line, the formula deflection (mm) = 246.5/(Young's modulus (GPa)) was obtained. This expression was then used to enable sufficient points to be plotted for the graph in Fig. 9.7(c) to be drawn.

Suppose that the graph of deflection against 1/Young's modulus had not been a straight line through the origin. It would have been possible, but tedious, to search for a straight-line relationship by plotting deflection against 1/(Young's modulus)^2, 1/(Young's modulus)^3 and so on. Sometimes dimensional analysis can be used to find the correct index – but it cannot help if, for example, the variables themselves are dimensionless numbers.

In many such cases, plotting with logarithmic scales can help. When two variables x and y are related by a power law such as

$$y = Cx^n,$$

then, taking logarithms of both sides,

$$\log y = \log C + n \log x.$$

It follows that, if the logarithm of y is plotted against the logarithm of x then the graph will be a straight line whose slope is equal to the n. To avoid having to look up a series of logarithms it is normal to use paper with logarithmic scales on which the lines of the reference grid are spaced in a similar manner to markings on a slide-rule scale.

An example of the use of such a plotting technique occurred during the tests on the structure described earlier. Four structures, each differing only in the one dimension, D, were tested with a single value of load and the deflection was measured in each case. The corresponding dimensions and deflections were plotted as in Fig. 9.8(a). The four points were then plotted on logarithmic scales, as shown in Fig. 9.8(b). They were found to lie on a straight line whose slope was measured as $a/b = -2.93$. Since most engineering systems are found to have simple values for exponents in governing equations, the value of n was taken as being exactly minus three. The data were plotted once more but this time as a graph of deflection against $1/D^3$ (see Fig. 9.8(c)) and were found to lie on a straight line through the origin indicating that

$$\text{deflection} = C/D^3.$$

The value of the constant, C, was found by measuring the slope of the graph. C could, of course, have been obtained directly from the log–log graphs but it would have been difficult to read the scales with sufficient precision. The slope was calculated from the actual dimensions of the graph measured with a rule. Reading values from the scales would have given the slope of the corresponding chord on the original curve.

Paper with logarithmic scales has many advantages in situations such as this but it can be infuriatingly confusing until its properties are familiar. It is necessary to have available a number of styles of paper. For example, if the range of deflection in Fig. 9.8(*b*) had been twice as big, it would not have been possible to accommodate the graph on the sheet by simply halving the vertical scale; instead the paper with a range of three decades would have had to be replaced with one having six decades.

9.13 Integration

Integration of test data can be performed graphically, numerically or, occasionally, analytically. For example, in order to find the volume of oil flowing through a circular pipe, a probe was introduced which measured the local velocity. As the skin friction causes the fluid velocity to vary from

Fig. 9.8. (*a*) Data thought to be governed by a power of the form deflection = CD^n. It is not possible to obtain the values of A and n directly from this plotted curve. (*b*) The data replotted using logarithmic scales. The points are found to lie on a straight line which confirms that they are governed by a power law. Measurement of the physical slope of the line gives the index in the mathematical relationship which is $n = a/b = -2.93$, say -3 in this case. (*c*) The data plotted a third time with scales shown to give a straight line relationship. The deflection has been plotted against $D^n = D^{-3}$. The slope of the resulting straight line is the constant C in the expression deflection = CD^n.

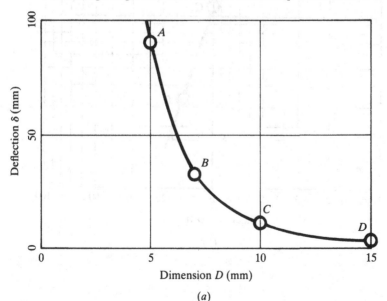

(*a*)

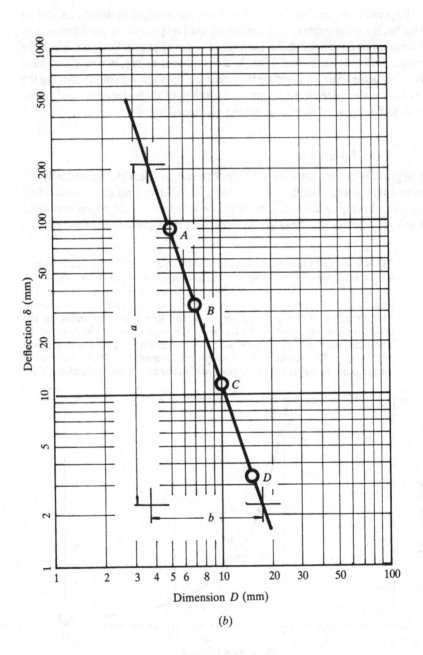

(b)

zero at the surface to a maximum at the centre, measurements of the velocity v were made at each of a number of radii, r. The readings are shown in Fig. 9.9(*a*). By considering the volume flowing in an annular element of radius r and thickness dr the total volume flow can be shown to be given by

$$Q = \int_0^{r_0} 2\pi r v \, dr,$$

where r_0 is the radius of the pipe which was 25 mm in the case considered. The values of rv were calculated and plotted as shown in Fig. 9.9(*b*). The volume flowing per second, which is equal to twice the area under this second graph, can now be obtained in a number of different ways.

BY COUNTING SQUARES

This elementary method should not be dismissed as childish. The technique is to count the number of squares of the reference grid lying beneath the curve. Most squares can be disposed of in blocks of 100 or 10 000, leaving those near the curve to be counted individually. Squares broken by the line are rounded up or down according to whether it is judged that more or less than half lie beneath the curve. A refinement to reduce the effect of human

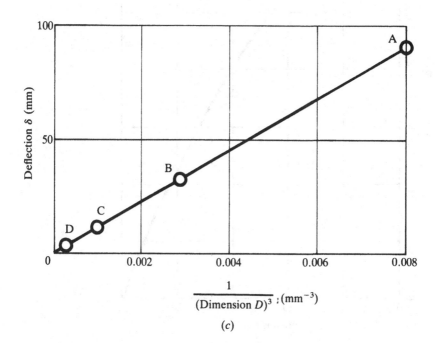

$$\frac{1}{(\text{Dimension } D)^3} : (\text{mm}^{-3})$$

(*c*)

error in this judgement and to reduce line thickness effect is to count up groups of squares alternatively above and below the line. The method is shown in Fig. 9.9(c).

Fig. 9.9. (a) The distribution of oil velocity in a circular pipe. To find the volume flow rate, a mean velocity is required. This mean velocity cannot be obtained directly from this graph. (b) The velocity measurements replotted to allow the volume flow rate to be found. The flow rate is proportional to the area under this curve which can be measured in a number of ways. The interpolated points were scaled off the graph to allow an accurate numerical integration to be made. (c) Integration by counting squares. Each shaded square counts as a whole unit. The unshaded squares count as zero. (d) The oil velocities replotted with scales chosen to give a straight line. It is now easy to find a mathematical relationship linking velocity and radius. The opportunity has been taken to 'non-dimensionalise' the graph to allow the findings to be applied generally.

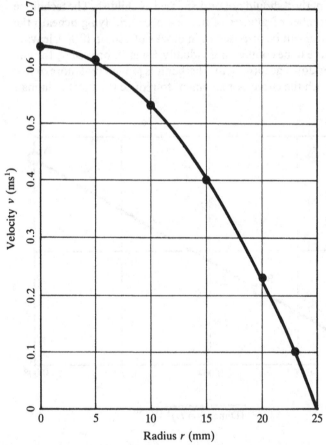

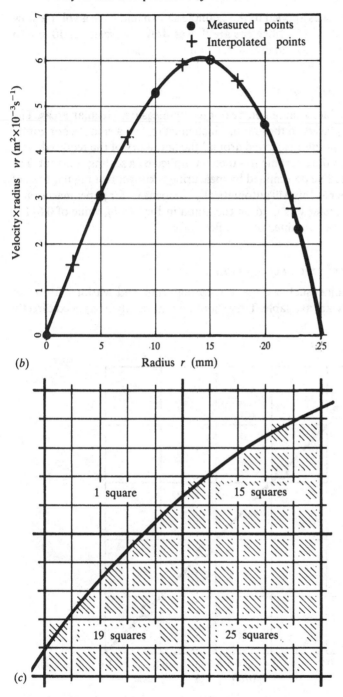

(b)

(c)

This method was applied to the example shown and gave a volume flow of 0.615×10^{-3} m^3 s^{-1} when the graph was drawn to scales of 10 mm to 0.5×10^{-3} m^2 s^{-1} and 10 mm to 0.4 mm.

BY PLANIMETER

If one is available, a planimeter can quickly measure irregular areas. The instrument is placed on the graph, which must rest on a smooth horizontal surface, and a pointer is moved around the boundary of the required area. The reading on the instrument is then multiplied by a scaling constant. The constant should be determined by measuring a known rectangular area of the graph paper as this will automatically compensate for paper inaccuracy. When a planimeter was used on the graph in Fig. 9.9(*b*), value of 0.621×10^{-3} m^3 s^{-1} was obtained for the flow rate.

BY NUMERICAL CALCULATION

Several formulae such as the trapezoidal rule and various forms of Simpson's rule are available. They should not normally be applied directly

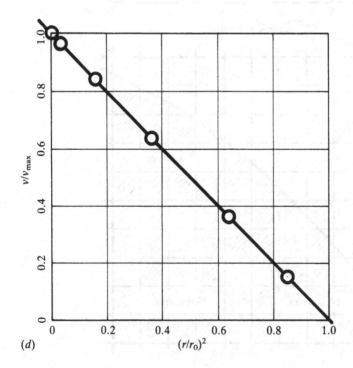

(*d*)

to measured data as they cannot make any allowances for scatter and the occasional misreading of an instrument. When applied to the pipe flow problem, Simpson's rule yielded a volume flow of 0.624×10^{-3} m³ s⁻¹.

In an attempt to improve the accuracy, the calculation was repeated using additional points interpolated from the graph. A new value of 0.615×10^{-3} m³ s⁻¹ was obtained.

BY CURVE FITTING

If an algebraic expression can be found to fit the experimental data, an algebraic expression for the integral will normally follow. Not only does this give a quick value for the integral but the solution is then available to be applied to integrate other, similar, sets of data with very little extra work.

In the particular case being discussed, it appeared that the 'velocity profile' was parabolic. Accordingly the graph shown in Fig. 9.9(d) was plotted of velocity against radius squared. To keep the solution general, both the variables were rendered non-dimensional by dividing by their respective maximum values. The straight-line graph shown was fitted by the equation

$$v/v_{max} = 1 - (r/r_0)^2;$$

from this expression the volume flow was calculated to be equal to $(v_{max}r_0^2)/2$ which equalled 0.618×10^{-3} m³ s⁻¹ in the case under consideration.

9.14 Interpolation and extrapolation

It is often found that measurements cannot be made with the apparatus operating at the exact conditions for which the results are sought. This might be because the variable which should be set to a particular value has to be regarded as a dependent variable for the purpose of the experiment. For example, to find the position of a machine part at a given time would be difficult, while it might be easy to measure the time at which it reaches a series of fixed positions. Again the independent variable might not be infinitely adjustable – a force might be applied by means of dead weights which increase by inconveniently large increments. A third possibility is that in the experiment the range of values available for a variable might be too small. For example, decisions involving the long-term strength or corrosion resistance of a material might have to be based on a relatively short test.

In all these cases the conditions at the required point can be read from a graph drawn through the available data. When the required point lies inside the range of plotted values the process is known as interpolation, otherwise it is extrapolation. Of course, there is no need to draw a graph in every case, mathematical methods, i.e. fitting an equation to the data, can be used.

INTERPOLATION

Probably the easiest method is to draw a graph. The scales should be chosen to give a fairly flat curve or, better still, a straight line. A false origin might improve accuracy as in the example illustrated in Fig. 9.3. As always, the common mistakes of graph drawing which were discussed previously must be avoided.

It is just possible that an unfortunate change in the character of the curve might happen between the plotted points. In Fig. 9.10(*a*) the measured amplitude of vibration of a motor car body at four particular frequencies is plotted. From the curve drawn through the experimental data the amplitude corresponding to a frequency of 2.5 Hz appears to be 6 mm. The real value at this frequency is nearly four times greater (21 mm), as can be seen from Fig. 9.10(*b*), where the same data has been plotted together with additional values at other frequencies. This phenomenon (resonance) is well known in both electrical and mechanical vibrations and the experimenter would normally be on the look out for it. In less-well-known cases similar behaviour can be detected by careful observation of the instruments and the apparatus while the variables are being changed. Where possible the measured values should be taken close to the interpolated points.

EXTRAPOLATION

Extrapolation is always less certain. It inevitably involves venturing into unexplored regions, so extreme caution is necessary. Consider, for example, the measurements illustrated in Fig. 9.11. Three pairs of values of the variables M and m lie on a straight line ABC. It seems reasonable to extrapolate this line a small way to the point D. The question is, what is the limit of such extrapolation? Is it also reasonable to extrapolate to point E and beyond? It is necessary once again to consider the physical situation. In this case M is the mass of a certain baby in kilogrammes and m is her age in months. Experience shows that in similar situations the mass does not increase linearly with time for long. It is unlikely that the child would weigh 27 kg (60 lbm) at the age of four years as represented by point E and unthinkable that after 20 years ($m = 240$) she would weigh 123 kg (272 lbm). Fortunately it is not necessary to have experience of similar cases to avoid the mistake of extrapolating to point E. There is a golden rule of extrapolation which says that: 'Extrapolation of a curve is not to be trusted when it is interpolation between an established point and the point $X = Y =$ infinity.' In a real physical system the mode of behaviour is likely to change at some point along the curve. Extrapolation right up to the point where

Fig. 9.10. (*a*) An illustration of the difficulties of interpolation. The four data points seem to lie on a smooth curve and it would appear reasonable to interpolate points such as 6 mm amplitude at a frequency of 2.5 Hz. (*b*) The correct shape for the previous curve. Extra data points show that the original curve was quite incorrect. Curves of this form are expected in vibration studies and the risks of interpolation are well known. Similar problems can occur unexpectedly in less familiar fields.

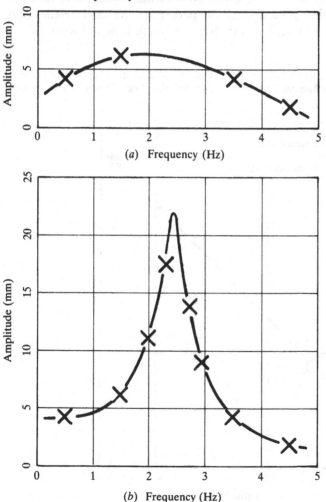

one variable (say, X) is infinite is quite in order but is difficult to draw. It is better to rearrange the variable as $1/X$ and extrapolate to zero.

9.15   Inspiration

So far only routine analysis has been considered. This is sufficient for nearly all experiments and the techniques can be tested and improved by practice. For a small proportion of experiments, something more is required. This extra is much more elusive and is difficult, some would say impossible, to teach. Every experimenter should be ready for the occasion when the results fall into a new pattern. It might be that experimental scatter is seen to be the effect of a previously ignored variable or perhaps a quite simple relationship can be spotted between two variables. Although little is known of the

Fig. 9.11. Extrapolation of data. The points A, B and C appear to lie on a straight line. It is reasonable to extrapolate this data to point D. What about point E?

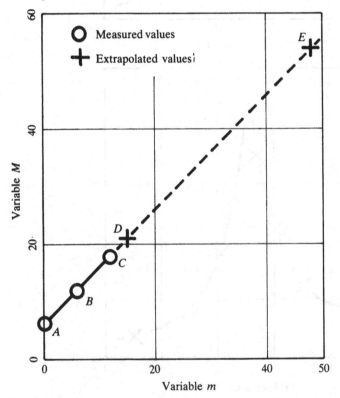

reason why sometimes 'things click into place', it is known that waiting patiently for them to do so is not sufficient. Waiting, for a day or so, can help but only after a problem has been considered and worried over in detail for long enough for the subconscious mind to take hold of it. Then stimuli from other problems or everyday living can be enough to give a new insight or a different viewpoint. For example, a highly successful altitude measuring device for aircraft is reputed to have been inspired by a pair of spotlights in a theatre. The inspiration only came because the inventor, before relaxing, had already spend a lot of effort seeking a way of measuring altitude.

The subconscious mind can be helped in its task if a conscious effort is made to look at a problem differently. A situation can be described accurately but differently by changing the viewpoint. A pessimist's half-empty wine bottle will appear half-full to the optimist. One person's careful judgement will seem guesswork to another person. It is sometimes useful to look at a situation from the directly opposite point of view to the obvious one. During the search for a suitable material for the light-sensitive drum in the Xerox photo-copier, selenium was tried and rejected as insufficiently sensitive. Then it was realised that it was too sensitive for the experiment; the stray light had 'fogged out' the image. A 'snag' which spoils a reading could be regarded as a fortuitous demonstration of a particular effect instead of a confounded nuisance. A spot of mould which spoiled a culture of bacteria led Fleming to the discovery of penicillin.

10
Communication of technical information

10.1 The need for effective communication

A very important task remains to be done after the experiments have been planned and completed and the results have been analysed and understood. This final stage is to explain what has been learned from the work so that other people can make use of it.

This stage deserves, but often fails to get, the same careful attention as the other aspects of experimental work. It is natural for the engineer's enthusiasm to fade when the satisfaction of solving problems and the excitement of making new discoveries are over; but unless his new-found knowledge is shared with others, all the effort spent on the planning and conduct of his experiments will be wasted.

Reports and lectures are the advertisements which make known the work done in the laboratory. Unless these advertisements are clear and attractive the engineer or scientist will not receive any credit for the work.

There are many other occasions when an engineer has to arrange and communicate information. He might have to explain new ideas or to persuade someone to agree to particular proposals. In fact, from first job interview right through to retirement speech, he is likely to spend more of his time communicating than calculating. So an important part of his training is the development of skills in both written and oral communication.

Some people have a natural flair for communication; most can be trained to a useful standard. The training includes practice and acquiring the habit of thinking in advance what effect each sentence, paragraph and illustration will have on the reader or audience.

280

To be successful, a communication has to hold the attention of the reader or listener against a host of unavoidable distractions. It must be short and interesting but must also convey the desired information clearly and completely. Communications can be either written or spoken. Some techniques are common to both methods but each has its own problems which will be considered separately.

Producing a report should not be regarded as an end in itself. Every report should have a definite purpose which must be understood before writing is started. This purpose, which is to transmit certain information to a particular group of people, will decide the form which the report takes. Many bad reports are produced by trying to write them in a standard form laid down by a particular organisation. The quality of a report should be measured by how well it transmits information and not by how closely it fits some arbitrary formula such as 'a picture, two tables, three graphs and 750 words'.

10.2 Preliminary ideas

When preparing a report or lecture, it is vital to visualise the attitudes and the competence of the people who will receive it. Reports on a single topic appearing in *The Engineer* and in *The Guardian* should be quite different even if they are written by the same person. A student would give different descriptions of his motor cycle to his colleague and his mother. Different people have different interests and experiences and these should be borne in mind by a writer or speaker.

Sometimes a single report must serve more than one group of people. Here special care is needed or none will find the report satisfactory. Normally one group has to be given priority but the others cannot be ignored.

Student laboratory reports are a special case. Although read and marked by someone with extensive experience of the relevant subject, they should be written as though intended for a reader at about the same level of technical training as the student. Another peculiarity of students' reports is that normally it is only the preliminary work which is submitted for marking – the final report is not produced.

When someone has been concentrating on a particular problem for some time, he might not appreciate which points will appear obscure to a newcomer to the topic. Every student will remember lessons which seemed completely meaningless because the lecturer had not remembered to explain an 'obvious' detail or to give enough background information. Being aware of the possibility of missing out a vital piece of information should prevent it happening. To make sure, a friend, or better still a colleague, should be persuaded to read through the draft report with a critical eye.

10.3   Guidance on the arrangement of written reports

Certain conventions must always be observed. These are flexible enough to leave room for some individual variations of style and presentation. The subject matter should be arranged in the following 13 sections.

> Title and author's name
> Summary
> Contents
> Notation
> Introduction
> Theory
> Experimental arrangements
> Results
> Discussion
> Conclusions
> Acknowledgements
> Appendices
> References

These groupings, which are explained below, will be familiar to the reader who will therefore be able to find and study a particular item without searching through the whole report. By arranging the work into labelled sections, the author will clarify his thoughts and be unlikely to omit anything of importance. The complete list shown is suitable for reporting on investigations lasting many months. For less ambitious projects, including student laboratory experiments, some simplification is necessary. The normal order of the sections is shown but authors might have to adopt a different arrangement to suit the house rules of their particular organisations.

Examples of reports on a range of subjects can be found in the proceedings of learned societies. These should be studied to see how experienced writers arrange their work. However, not every published paper should be regarded as a perfect example to imitate; students should look critically at papers and articles to see why some are easier to understand than others.

The length of a section might be anything between zero and many pages depending on how much work is being reported. Information should only be included where it will be of interest to the intended reader. In long reports, the sections should be divided into sub-sections which should be given sub-titles.

In their early years of study, students will normally be told how much detail to put in their reports. To save work they might be asked to omit certain sections entirely. Frequently no description of the equipment or method of operation is required. Such incomplete reports allow the demonstrator, who is already familiar with the experiment, to check quickly whether a standard exercise has been completed properly and understood. They are not meant to be read by a newcomer to the experiment.

The contents of the different sections are:

§1 TITLE AND AUTHOR'S NAME

The title should be short but should convey a clear idea of the subject matter. For example, 'Cyclic Pressure Tests on Glass Reinforced Plastic Pipes' rather than either 'Pipe Test' or 'Measurement of the Fatigue Life of Twelve Samples of 300 mm Diameter Glass Reinforced Plastic Water Pipes when Subjected to Slowly Applied Pressure Cycles in the Range 0 to 2.5 MPa'.

§2 SUMMARY

In this section the purpose and extent of the report are outlined and the main conclusions are given very briefly. The reasons for doing the work should be clear, but they should not be given in a bald statement of a target. For example, it is better to say: 'The thermal conductivity of PVC was measured. . . .' than 'Object: to measure the thermal conductivity of PVC.' This convention marks an important difference between technical reports and the 'write ups' for school laboratory experiments.

The summary is designed to enable a reader to decide quickly whether the report contains anything valuable in his field of interest. It should be worded to help him, not to trap him into wasting time on a report which is of no value to him. The summary is normally printed at the beginning of a report; but it should be written last. When deciding what to put in it, the author should ask himself the question: 'What did I manage to achieve in my work?'

There is no need to give this paragraph the title 'Summary'. Its position between the main title and the body of the report is sufficient to identify its purpose.

§3   CONTENTS

A list of contents, giving the page number of each section or sub-section and each illustration, is required in long reports. The contents should be listed only when it is expected to be of real assistance to the reader and not just to make a report seem more formal.

§4   NOTATION

This is a list of brief definitions, with units where appropriate, of every symbol appearing in the report. The symbols are arranged in alphabetical order, English letters first, followed by Greek. Some authors also define each symbol when it first appears in the body of the report but it is normally sufficient to do so just for any unusual symbols.

Some recommended symbols are shown in Table 2 of Appendix 3. A more comprehensive list is given in BS 1991 (Ref. [1]).

§5   INTRODUCTION

This section puts the work which is being reported into perspective. It starts with a brief review of previous work on the same topic and gives references to existing publications. Then it explains why and how the current work was done. Finally the range of the work is outlined.

§6   THEORY

The relevant theory, if there is any, should be given, but only in outline. Each basic equation should be introduced with an explanation, where appropriate, of its physical significance. Any derived equations or solutions should then be written with a brief description of the method of manipulation, but without giving detailed intermediate steps. Sometimes, even in outline form, the theory will be excessively long. In such a case, or when a lot of detail is required to explain an unusual method, it is sensible to give just the barest outline at this stage and to put the full theory in an appendix at the end of the report. Once more the author must bear in mind his typical reader's capability and experience. He must also help a newcomer to the particular topic by giving references to the previously published theories on to which his own work was built.

§7   EXPERIMENTAL ARRANGEMENTS

The equipment and experimental procedure should be described. Detailed descriptions of novel or complicated equipment should be given in a further appendix. Carefully prepared line drawings and, occasionally, photographs should be included where they will help the reader to understand the arrangement. A report should not normally contain a list of every item of equipment used or of identification numbers of measuring equipment. This information, which should, of course, be recorded in a laboratory log book, will not interest the normal reader.

§8 RESULTS

The results obtained from both theory and experiment should be given next. Frequently a graphical presentation is the easiest to understand and enables the differences between calculated and measured data to be seen. It is not normally necessary to give results in both numerical and graphical forms, but where an expensive or complicated experiment has been performed to obtain basic physical properties, the actual readings taken should be included in the report. Experimental measurements will anyway be preserved in the laboratory log book.

The units must be given for all physical quantities. The units used should be those normally accepted and therefore familiar to the reader. They should be consistent throughout the report. Where a report is written for readers who use both SI and Imperial or US Customary units, then one system should be chosen as the main one but values in the other system should be added in brackets, e.g. '0.97 kg (2.1 lbm)'. As well as measured and calculated data, this section should contain descriptions, with photographs and line drawings, where helpful, of such results as flow patterns and fracture surfaces.

§9 DISCUSSION

The author's own comments on his work appear here. He might wish to point out strengths and weaknesses of his experimental arrangements or of his method of calculation. He should compare his method and results with those of other workers. He will nearly always have something to write about the accuracy of his data.

The significance of the results should be discussed. For example, there might be a good or a bad agreement between the experiment and a given theory. Actually, outside schools and universities, few reports are concerned with confirming theories; the conclusions are more likely to be about the suitability of a particular design or material for a certain duty. An important possibility to bear in mind is that a set of experiments might give inconclusive results not capable of either confirming or disproving a particular theory or perhaps not allowing a decision to be reached about the likely life of a certain component.

Since all experimental and theoretical investigations are part of a continuous unfolding of new knowledge, each report should illuminate the way a little further. Suggestions for further work in the same field should be made. Occasionally, it can be stated that a particular line of enquiry is no longer worth pursuing. Further work which the author has already started should be mentioned to reduce the chances of unnecessary duplication.

§ 10   CONCLUSIONS

In this section the main results are summarised. This section should be kept short enough for the main conclusions to be found quickly by a casual reader. It should not be a further discussion about the work done. By reading this section after the title and the summary, a reader should get a clear idea of the scope and limitations of the whole report.

§ 11   ACKNOWLEDGEMENTS

As a matter of courtesy, thanks should be expressed for any assistance in the form of facilities or advice. This custom has the advantage of making known departments and people who have developed particular skills in auxiliary fields as well as giving credit where it is due.

§ 12   APPENDICES

As already mentioned, detailed theory and descriptions of apparatus should be appended at the end of the main body of the report. Most readers will not wish to study these in detail, at least during their first reading of the report. The use of appendices is especially valuable when two classes of reader are being served. For example, the main body of the report might be aimed at a general reader who is mainly interested in the conclusions, while appendices satisfy the needs of someone doing similar work and having a more detailed interest in the methods used.

§ 13   REFERENCES

When reference is made to a published work in the body of the report, a reference number is inserted into the text. These references are listed at the end of the report. The standard form for each reference is: 'Number, author's name and initial, title of paper, title of journal or book, publisher, volume number, page number, year of publication'. The reference should be given in the order of first appearance rather than in alphabetical order. Examples of the format used for giving references can be found throughout this book.

10.4 Standards of writing and illustrations

It is normal to write technical reports in the third person, past tense, e.g. 'the diameter was measured' rather than 'I measured the diameter'. This practice results in impersonal, modest reports and avoids difficulties when reporting the combined operations of a group. There are occasions when other styles might be more suitable. Whichever style is chosen should be maintained consistently throughout the report.

The use of simple words in simple sentences will make a report pleasant to read. Several 'fog' or 'clarity' indices have been suggested. These are claimed to give a measure of the readability of a report. They are normally based on the average number of words in each sentence (wps) and the percentage of words containing more than two syllables (pop). To make it easy to do the analysis by computer, a polysyllable is usually interpreted as a word with more than eight letters. An example of the type of expression used is:

$$\text{fog index} = (\text{wps} + \text{pop} + 3)/5.$$

This index lies between six and eight in a well-written report. Of course, it is not yet possible to assess the quality of reports completely by such formulae. Many other factors affect the readability and clarity.

The subject matter, not the style of writing, should hold the reader's attention. Superlatives such as 'greatest', 'best' and 'most' should be reserved for maxima in the mathematical sense. Where possible, clichés and catch phrases should be avoided. Advice on both style and acceptable grammar can be obtained from any of a number of standard books; for example, Fowler (Ref. [2]) and Gowers & Fraser (Ref. [3]). Engineering students should obtain and study at least one of these books. They are not dry and dusty collections of outdated rules, but rather interesting guides to the production of clear and readable work.

It is often convenient to abbreviate frequently recurring phrases. Unless the abbreviation is likely to be familiar to the intended reader, the first time it occurs the phrase should be given in full with the abbreviation added in brackets. For example, early in a report the phrase '. . . a complementary metal oxide semiconductor (CMOS) decoding chip . . .' might be written. Then for the remainder of the report the abbreviation 'CMOS' would be used without further explanation.

Single words should not normally be abbreviated – it is not acceptable to write 'gas turb.' instead of 'gas turbine' or 'lab.' for 'laboratory'. Common exceptions to this rule are 'Fig.', 'Eq.' and 'Ref.' for 'Figure', 'Equation' and 'Reference' respectively. Colloquial abbreviations such as 'can't' and 'shouldn't' are not used in technical reports.

Whole numbers up to ten, or perhaps twelve, should be written as words, above that as numbers, e.g. 'seven' and '17'. However, it is preferable that the way of writing numbers should not be mixed within a single sentence. For example, 'between 6 and 13' is better than 'between six and 13'.

It is sensible to use a single series of numbers, i.e. Fig. 1, Fig. 2, etc., to refer to all the illustrations. However, some authors and organisations use a different series for each type of illustration such as photographs, line drawings and graphs. This second method of labelling has no real advantage and makes the report discontinuous and less attractive. Tables of data are normally put in a different series from the other figures.

Drawings are useful in reports as they can transmit engineering information very clearly and economically. Before producing a drawing, the author must decide what information he wants to convey and what ability his reader is likely to have in interpreting drawings. Detail and assembly drawings, as used for manufacture, are not normally suitable for technical reports. For instance, in reading about some tests on the temperature distribution in a diesel engine piston, the reader should not be distracted by dimensions and tolerances of every feature of the piston. He should be given a general impression of the shape and size and the positions of the temperature measuring points.

Fig. 10.1. An illustration which does very little to help the reader.

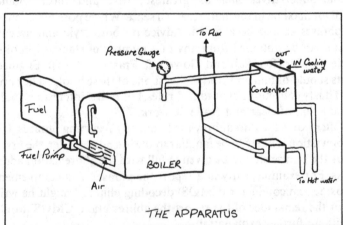

Pictorial views are sometimes thought to be simple to understand. However, they are difficult to produce and are not very efficient except for showing the relative positions of features of apparatus – information which is frequently unimportant. More useful are schematic diagrams; electronic circuit diagrams are the best-known example but similar techniques work well in other fields. For example, flow diagrams can illustrate chemical processes and computer programs. Whatever type of drawing is chosen, it should be produced in boldly ruled ink lines.

As an example to be avoided, a performance test on a small oil-fired boiler was illustrated by Fig. 10.1. An unskilled attempt to draw an isometric view gave a rough idea of the general appearance of the plant but its method of operation was not at all clear. A broader pen should have been used and the features should have been labelled in block letters. No reliance can be put on the relative sizes and positions of the parts as these might well have been distorted to fit everything in the diagram.

Fig. 10.2 represents the same boiler system. This diagram was slightly easier to draw and is much easier to understand. The principle of operation is now obvious and the quantities measured can be seen easily. Attention has been paid to the line thickness and the lettering. Finally, a meaningful title has been used.

Graphs are a very common way of showing results. Normally a reader will expect to obtain qualitative, rather than precise quantitative, information from them. So, in printed reports, only enough reference lines to

Fig. 10.2. A better drawing of the boiler shown in Fig. 10.1. This schematic view with careful lettering is much clearer than the first attempt but was not much more difficult to reproduce.

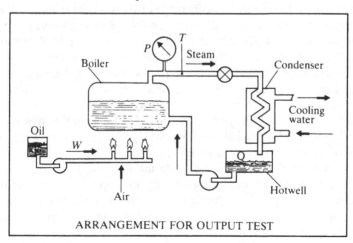

establish the scale are included. The curves should be drawn with thick lines to enable them to stand out against the background. The general shape of the curves can then be seen at a glance. (Graphs used during calculations and in data sheets have a different purpose and are drawn with fine lines on a detailed reference network.) For undergraduate reports, a similar effect can be produced quite easily. No tracing is required provided pale blue or green graph paper is used. The reference grid should be drawn on the graph and a photocopy made. The printed squares used for plotting will not appear on the copy. Some printers will accept graphs produced in this way; others will require an ink tracing.

When a graph is used to enable calculated and measured values to be compared, it is normal to make the curve represent the calculated values, and to show the measurements as individual points. A variety of symbols, e.g. crosses, squares and triangles, can be used to distinguish different sets of results.

Where a number of graphs of similar results are included in a report the same scale should be used for each to simplify comparisons. The scales should always be chosen to make each division of the reference grid equal to one, two, five or ten units of the variable rather than the less intelligible three, four, seven, etc., units. If a graph is drawn with a 'false origin' a note drawing the reader's attention to it will reduce the risk of misunderstanding. More detailed advice on drawing graphs is given in a booklet (Ref. [4]) issued by the American Society of Mechanical Engineers.

Fig. 10.3 shows a badly drawn graph which was drawn in an attempt to show how the results of a particular experiment on yielding under compound stresses compared with a theoretical prediction.

The scales were badly chosen; space is wasted on the horizontal axis and the divisions are awkward. It is difficult to distinguish the theoretical from the experimental values. Both the curve and the plotted points are far too feeble – they do not stand out clearly against the reference grid. Worst of all, the curve has been drawn freehand instead of with the help of French curves. The units of the horizontal axis 'stress $N/m^2 \times 10^6$' are ambiguous – they could be read as either (stress $\times 10^6$) in $N\,m^{-2}$ or stress in $(10^6 \times N\,m^{-2})$, i.e. in $MN\,m^{-2}$.

A better attempt at displaying the same data is shown in Fig. 10.4. As well as avoiding the faults listed above, a more helpful title has been added and attention has been paid to the lettering.

Four different methods are available for lettering graphs and other diagrams. In order of increasing quality of appearance and also of increasing difficulty of production they are: typing, hand lettering, stencilling and using dry transfers such as Letraset. One single method should be used throughout the report.

It should be noted that careful hand lettering is quite adequate for most engineering reports. The number of sizes and styles of lettering used should be restricted; it is rare for more than two sizes to be needed.

Sometimes when diagrams are being prepared for printing, all the lettering is left for the printer to insert. A second copy of each diagram with the wording pencilled in is included to show what is required.

The illustrations are distributed in a printed report so that each one is close to the relevant part of the text. This distribution is made by the printer; the author groups the text and the illustrations separately. For typewritten reports the illustrations are normally grouped together at the end of the report. There are two reasons for presenting the report this way. Firstly in a typed report it is difficult to arrange for both an illustration and its relevant text to be visible together, particularly when only one side of the paper is used. Secondly it is slow and difficult to type a report leaving gaps of the

Fig. 10.3. A graph which was drawn to show the results of an experiment to measure yield strength but which succeeds in showing all the common faults which are found in graphs.

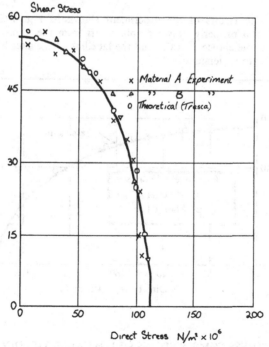

292   *The systematic experiment*

right size for illustrations in the right places and then to draw the diagrams in the gaps.

10.5   Building up a report

Even a highly experienced author cannot hope to write the first draft of a report in a single operation. Instead he has to build it up gradually by going through several stages. The process is similar to the way a designer builds up a drawing starting with a blank piece of paper and using feint lines, which are easy to erase, until the final shape is decided. It is most important to delay the actual writing of any complete sentences until the general form of the report has been finalised.

The first draft of a report will pass through the following nine stages:

§1   A list of the reasons for doing the work and for writing the report together with a note of the main characteristics of the intended reader.

§2   A list of chapters or sections.

§3   A list of items to be described and figures to be drawn in each

Fig. 10.4. The results of the experiment shown in Fig. 10.3, but here presented in a properly drawn graph. Apart from the improved general appearance, the choice of scales and the labelling of the axes have made it much easier to understand.

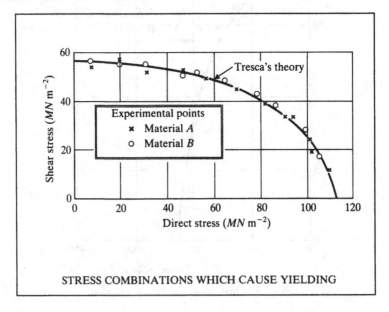

chapter or section. A couple of key words are sufficient to identify each item. The summary and the conclusions should be left until later.

§4 The same list but with the items in each section arranged into a logical order. By this stage some items might have been transferred to a more appropriate section.

§5 A check to make sure that sufficient information is available. The need for more calculations or experiments might be seen at this point.

§6 A check that the report as outlined will fulfil the stated objectives.

§7 The list of items in each section expanded into acceptable English with a balance between the lengths of the different parts.

§8 Rough drawings of the graphs and other illustrations with figure numbers and titles.

§9 The addition of the conclusions and, finally, the summary.

The rough version of the report which is produced by the sequence just described will have to be refined several times. First, a good balance of length must be achieved between the different sections. Then each sentence must be checked to make sure that it will convey the desired meaning to the person for whom it is written. Some words such as 'heat' and 'strain' have particular meanings for engineers which are slightly different from their everyday usage. Each word has to be selected carefully to perform a particular task. Just as the instruments used during an experiment should be chosen with care, so the words and phrases used in the report should be examined carefully to ensure that each one is the most suitable. No two words or phrases carry exactly the same meaning. 'A half' is not the same as '0.5', 'rapidly' is not interchangeable with 'quickly'.

Some groups of words can have a particular meaning which might not be the one intended. A simple experiment will show the different effect of these two similar sentences: 'When I look into your eyes, time seems to stand still," and 'You have a face that would stop a clock.' 'High speed engine test' can have two meanings. Hyphens can sometimes make the intended meaning clear, for example, 'high-speed engine test' would be a rapid test on an engine; 'high-speed-engine test' would mean a test on a high-speed engine.

Pronouns should only be used when it is quite clear what they mean. For example, in the sentence: 'The output from the thermocouple was fed to an amplifier which was connected to a recorder; unfortunately its time-lag led to inaccurate readings' – it is not clear which of three possible things had the time-lag. Any doubtful or ambiguous sentences should be rephrased.

Further reading of the report might reveal excessive repetition of a single word or phrase. For example, a number of sentences might be found to start with: 'Next this was done . . . .' If possible these irritants should be changed, but not at the expense of clarity.

A lot of time needs to be allowed for preparing a report. Repeated readings should be spaced out over a period of several weeks if possible, so that ideas have time to form between readings, and so that a fresh look is taken each time a draft is read. After the author himself has read the report several times and made any obvious improvements, he should ask a colleague to examine it. The colleague should be able to find any parts where the meaning is not clear or where too much prior knowledge has been expected of the reader.

It is becoming commonplace for reports to be prepared on a word processor. Where no professional word processor is available, use can be made of either a word processing program on a micro-computer or the file editing facility on a larger computer such as a VAX. The ability to make both structural changes and detailed corrections to a report without the labour of retyping or rewriting means that there is no reason for poor quality reports to be produced ever again.

Normally the first version is typed in professionally from dictation or manuscript. Then, unless major changes are necessary, the author makes the refinements. He can then see immediately how an alternative version of a sentence or paragraph would appear. Just as when editing computer programs, it is easier to work from a printed copy of the first draft rather than directly from the screen.

The temptation to use the 'justify right margin' facility which puts in extra spaces to straighten out the right-hand edge of the text should be resisted. The overall impression is similar to a printed page but, unless the printer used has both proportional and micro spacing, then the uneven gaps between the words will make the text very tiring to read.

Care is required to avoid filking. This is the process where an adequate sentence is altered in several stages until the final version is much worse than the original. The phenomenon is similar to the instability that is sometimes found during iterative solutions of equations. Filking can be avoided by allowing plenty of time between stages so that the work is always fresh.

One last necessity is a check on spelling. It must be realised that poor spelling will give an impression of slovenliness, not eccentricity, and will reduce confidence in the report. Poor spelling will cause considerable extra cost and delay when a manuscript is sent for printing. Where the report is to be typed, it is not the typist's duty to correct grammar and spelling; the total responsibility lies with the author. Poor spellers can be helped by the dictionaries which are built into some word processors but these still require knowledge and judgement from the user. For example, a spelling checker would pass both 'principle' and 'principal' as correct; the author would have to decide which spelling is right in the particular context.

Many people have difficulty producing the seventh stage of their first draft, that is, in expanding the jottings into sentences and paragraphs. The difficulty is caused by the realisation that a report is something special which will be read more critically and which will be more permanent than most of their writing. The feeling of nervousness is similar to that experienced by most drivers when they realise that there is a police car behind them. Of course more care has to be taken over the style and grammar than in everyday speech, but the language used should be basically the same.

One way of getting started is to say out loud whatever needs to be put in writing. Because speech is more familiar, the author will be more at ease and the words will flow freely. Once some words have been found to express the ideas, they can be written down, examined and modified if necessary.

10.6 Shorter communications

Many situations do not call for a full report as outlined above. Some sections might be shortened or omitted entirely since information should only be included when it is certain that the reader wants it. Frequently it is known that the reader will only require, for example, details of a particular apparatus or the results of a certain series of measurements. In such cases the required material should be arranged in a logical order and refined as though it were part of a full report. Care must be taken to ensure that sufficient information is included. For example, the power curves for an engine are of no use unless the ambient pressure and temperature are known.

The most common written communication is a letter. Here time is not normally available for rough drafts to be made and refined. All the information has to be assembled as a very rough list and sorted mentally during writing or dictation. Letter-writing ability will grow with practice provided a critical look is taken at each attempt with a view to doing better the next time.

10.7 Oral presentation of reports

Skill in speaking is required daily and not just for the occasional formal lecture, so engineering and science students should find ways of improving their ability. To be able to put a point of view or to explain an idea clearly and succinctly is a considerable advantage in most careers. The importance of skill at oral presentation is reflected in most university and college courses where a lot of marks depend on the student's presentation and defence of a thesis.

For informal discussions, either direct or by telephone, the technique is similar to that of letter writing. The relevant information must be assembled either mentally or as a written list of topics before speaking and no rough drafts are possible. Fortunately, the listener can ask for clarification of particular points. It is important to think before starting to speak; sometimes: 'It is better to be silent and let someone think you a fool than to speak and let him know for certain.'

Most formal lectures are presentations of reports which have already been written. In other cases it is necessary to collect and arrange the information as though preparing a written report. It has been found that the effect of reading a carefully composed report word for word is extremely boring and unnatural. Even worse is the effect of reciting a memorised speech. So, instead of a finished report, a detailed list of headings and topics should be taken to the lecture. This outline can then be filled in spontaneously during delivery.

An oral presentation has the considerable advantage of allowing the speaker to see immediately the effect he is having on his audience. This enables him to change his approach (or, in an extreme case, his job!).

Listening is a much slower way of receiving information than reading, so care is necessary to avoid trying to include too much in a talk. For example, a 'quality' Sunday newspaper which occupies most readers for a couple of hours would require about 40 hours to transmit by television. It follows that most of the detail given in a written report has to be omitted from a lecture. A speaking rate of about 100 words per minute should be assumed when planning to make a talk fit into a certain time. A simple illustration will take about a minute for the audience to absorb, a complicated one requiring a lot of explanation will need much longer.

It is well worthwhile rehearsing any important oral presentation. Even experienced speakers underestimate how long it takes to present a prepared talk. Many conferences and examinations are run with very strict time limits. When a talk is cut short by the chairman it is embarrassing for the speaker and frustrating for the audience.

The subject matter must be completely understood before starting, in order to give confidence. However, there is no need to memorise details of the order in which each topic will be explained. A set of notes should be used quite openly. These notes should comprise a list of all the topics which are to be explained. The speaker can then be relaxed knowing that nothing will be omitted or presented in the wrong order.

Formal speaking should be started early in a student's career when the audience will be tolerant and sympathetic. The professional institutions have sections for younger members where practice and advice are freely available. It is a good idea to give a talk and to have it recorded, if possible on video. The recording can then be heared or viewed several times in private. Defects in presentation such as mumbling, rushing or constant use of the same word or phrase will be obvious. Fortunately, because the audience hears a talk once only, slight errors in style are not normally noticed in practice.

Because the audience cannot stop to think or to refer back to a previous item, it is necessary to repeat the main topics. It is said that army instructors successfully obey the slogan: 'First you tells 'em what you're going to tell 'em. Then you tells 'em. Then you tells 'em what you've told 'em.'

Illustrations are very important during lectures. They should be prepared on slides or overhead projector transparencies and not handed round on paper. Only a small amount of information should be shown on each illustration. Experiment and experience will be necessary to determine the best line thickness and the most suitable size of lettering for the room and projector which are to be used. Illustrations will normally be drawn specially when making of transparencies for projection. It is unlikely that printed matter from books will be suitable, it will probably have too much fine detail.

With forethought, the speaker and audience can be allowed to concentrate on the subject matter of the talk without being distracted during the presentation. If at all possible, the projector should be loaded and focused before the audience enter the room. At the same time the pointer should be tried and put ready to hand. A title illustration left projected while the audience assemble will ensure that they settle where they can see the screen instead of having to move around after the talk has started. When slides are being used and a particular illustration needs to be shown twice, two copies should be made. The slides can then be presented in a smooth sequence instead of skipping backwards and forwards.

If this chapter has made successful communication appear to be hard work, then it has conveyed the correct impression. There are no short cuts, but the effort to succeed is well worth making. The young engineer should learn the art in order to make the most advantageous use of his traditional training throughout his career.

References

[1]   BS 1991: Part 1, *Letter Symbols, Signs and Abbreviations*, 1967.

[2]   Fowler, H. W. & Gowers, Sir Ernest (1981). *Modern English Usage*. Oxford University Press.
[3]   Gowers, Sir Ernest, & Fraser, Sir Bruce (1977). *The Complete Plain Words*. HMSO.
[4]   American Society of Mechanical Engineers (1964). *An A.S.M.E. Paper*. Manual MS-4, April 1964.

Appendix 1
Outline of dimensional analysis

A1.1 The principle of dimensions

The basis of dimensional analysis and its logical development has been set forth elsewhere (Ref. [4.2]). The full logic introduces, at one stage, the idea that dimensions, like physical quantities, can be formed into product groups enabling the idea of cancellation of dimensions to be introduced. For example, the quantity, p/ρ, which appears in the Bernoulli equation for flow of a fluid has dimensions

$$\frac{ML^{-1}T^{-2}}{ML^{-3}}=\frac{L^2}{T^2},$$

here, cancellation of both mass and length dimensions has been done.

At another stage in the logic the idea of the equality of dimensions is introduced. An example is provided by the Bernoulli equation for a flow of fluid, which is

$$P/\rho+gz+\tfrac{1}{2}V^2=K. \tag{A1.1}$$

Using the $M, L, T,$ system of denoting the dimensions of mass, length and time, respectively, the dimensions of gz are

$$LT^{-2}L=L^2T^{-2},$$

and the dimensions of $\tfrac{1}{2}V^2$ are

$$(LT^{-1})^2=L^2T^{-2},$$

and the dimensions of P/ρ have already been quoted above as (L^2T^{-2}).

299

The dimensions of all three terms of the Bernoulli equation are seen to be identical and, further, K must have the same dimensions. Hence a change in the system of units changes each term by an identical factor whilst the form of the equation is unchanged.

Equation (A1.1) can be put in a non-dimensional form. Just one arrangement is

$$\frac{p}{\rho K}+\frac{gz}{K}+\frac{1}{2}\frac{V^2}{K}=1,$$

(A1.2)

each one of the three terms on the left-hand side being dimensionless as is the right-hand side.

A1.2   The pi theorem

The most general form of equation (A1.1) is

$$f_1(p, \rho, z, g, V, K)=0,$$

(A1.3)

the form of the function f_1 being unspecified. In equation (A1.3) there are six variables and they require three dimensions to fix their numerical values, these dimensions being M, L and T.

The pi theorem shows that these six variables can be re-arranged into non-dimensional groups. The procedure (Ref. [4.2]) in this case is now set out.

Firstly, the variables and their dimensions are set out as follows:

(1)   Variable p ρ g z V K;

(2)   Dimension $\dfrac{M}{LT^2}$ $\dfrac{M}{L^3}$ $\dfrac{L}{T^2}$ L $\dfrac{L}{T}$ $\dfrac{L^2}{T^2}.$

The quantity ρ is now used to cancel the mass dimension. The result is again set out as follows:

(3)   Variable $\dfrac{p}{\rho}$ ρ g z V K;

(4)   Dimension $\dfrac{L^2}{T^2}$ $\dfrac{M}{L^3}$ $\dfrac{L}{T^2}$ L $\dfrac{L}{T}$ $\dfrac{L^2}{T^2}.$

Equation (A1.3) can now be written as

$$f_2\left(\frac{p}{\rho}\cdot\rho,\rho,g,z,V,K\right)=0$$

or

$$f_3\left(\frac{p}{\rho},\rho,g,z,V,K\right)=0$$

(A1.4)

Inspection of line (4) above shows that ρ is now the only variable in equation (A1.4) that contains a dimension of mass. The argument now is that if there is to be an equality of dimensions throughout this equation so that the mass dimension could be cancelled out then so also must the variable, ρ, cancel out numerically from this equation so that it reduces to:

$$f_4\left(\frac{p}{\rho}, g, z, V, K\right) = 0. \tag{A1.5}$$

Again, these variables are tabulated as follows:

(5) Variable $\dfrac{p}{\rho}$ g z V K;

(6) Dimension $\dfrac{L^2}{T^2}$ $\dfrac{L}{T^2}$ L $\dfrac{L}{T}$ $\dfrac{L^2}{T^2}$.

Now the variable K is used to cancel the time dimension. The result gives a listing as follows:

(7) Variable $\dfrac{p}{\rho K}$ $\dfrac{g}{K}$ z $\dfrac{V}{\sqrt{K}}$ K;

(8) Dimension 1 $\dfrac{1}{L}$ L 1 $\dfrac{L^2}{T^2}$.

Now equation (A1.5) can be written as:

$$f_5\left(\frac{p}{\rho K}\cdot K, \frac{g}{K}\cdot K, z, \frac{V}{\sqrt{K}}\cdot \sqrt{K}, K\right) = 0$$

or $\qquad\qquad$ (A1.6)

$$f_6\left(\frac{p}{\rho K}, \frac{g}{K}, z, \frac{V}{\sqrt{K}}, K\right) = 0$$

As previously argued, the variable K being the only one in equation (A1.6) that contains a dimension of time, it can be cancelled out so that

$$f_7\left(\frac{p}{\rho K}, \frac{g}{K}, z, \frac{V}{\sqrt{K}}\right) = 0. \tag{A1.7}$$

Tabulating again gives:

(9) Variable $\dfrac{p}{\rho K}$ $\dfrac{g}{K}$ z $\dfrac{V}{\sqrt{K}}$;

(10) Dimension 1 $\dfrac{1}{L}$ L 1.

The variable, z, is now used to cancel the length dimension. The result gives the listing:

(11)   Variable $\dfrac{p}{\rho K}$ $\dfrac{gz}{K}$ z $\dfrac{V}{\sqrt{K}}$;

(12)   Dimension   1   1 L 1.

Thus equation (A1.7) is written as

$$f_8\left(\frac{p}{\rho K}, \frac{gz}{K}\cdot\frac{1}{z}, z, \frac{V}{\sqrt{K}}\right)=0$$

or   (A1.8)

$$f_9\left(\frac{p}{\rho K}, \frac{gz}{K}, z, \frac{V}{\sqrt{K}}\right)=0$$

Now the variable z is the only one in equation (A1.8) containing a dimension of length so that it is cancelled giving

$$f_{10}\left(\frac{p}{\rho K}, \frac{gz}{K}, \frac{V}{\sqrt{K}}\right)=0$$

or

$$f_{11}\left(\frac{p}{\rho K}, \frac{gz}{K}, \frac{V^2}{K}\right)=0.$$

This is the required result.

Once familiarity with the technique has been acquired then the tabulation can be set out as follows:

Variable	p	ρ	g	z	V	K
Dimension	$\dfrac{M}{LT^2}$	$\dfrac{M}{L^3}$	$\dfrac{L}{T^2}$	L	$\dfrac{L}{T}$	$\dfrac{L^2}{T^2}$

Variable	$\dfrac{p}{\rho}$	g	z	V	K
Dimension	$\dfrac{L^2}{T^2}$	$\dfrac{L}{T^2}$	L	$\dfrac{L}{T}$	$\dfrac{L^2}{T^2}$

Variable	$\dfrac{p}{\rho K}$	$\dfrac{g}{K}$	z	$\dfrac{V}{\sqrt{K}}$
Dimension	1	$\dfrac{1}{L}$	L	1

Variable	$\dfrac{p}{\rho K}$	$\dfrac{gz}{K}$	$\dfrac{V}{\sqrt{K}}$
Dimension	1	1	1

Appendix 2
Outline of error analysis

A2.1 Introduction

As it was stated in Chapter 8, the purpose of this appendix is to describe some elementary techniques of evaluating experimental observations affected by errors and other random factors. It is not intended here to embark on an extensive review of statistical methods, but merely to outline, and to demonstrate the use of a few of the most essential procedures and formulae furnishing them with some fundamental definitions and statements. Several illustrative examples are interjected individually in places with the intention of showing how the immediately preceding material can be applied in real situations. One entire section, namely A2.7, constitutes nothing else than just one single exercise covering the material presented in all previous sections. This should be instrumental as processing of a more or less complete experimental case is considered there.

Imagine a sequence of measurements or (speaking in more general terms) observations whose nominal values are expected to remain constant. In fact the observations, as pointed out in Chapter 8, will most likely exhibit a random experimental scatter. For example, the actual numerical values of a mechanical property of material (e.g. yield stress) may vary from specimen to specimen, even though the specimens are supposed to be technically identical, i.e. made from the same material and to the same manufacturing specifications. Similarly, a series of identical automobile engines when tested on a testing rig are likely to produce somewhat different performance parameters (e.g. fuel consumption, maximum power, torque, etc.) for the same test conditions. Also, lives of individual electronic transistors, from the same production batch, will not be exactly equal. In all these cases the

303

actual items whose performance is being monitored may be slightly different and there may be experimental errors in the relevant observations. Both components are inherently random, i.e. without a coherent pattern. However, in spite of that, a rational quantification of such randomly scattered observations is possible, thanks to several simple statistical rules which will be discussed below.

A2.2   Presentation of observations

Any set of randomly scattered observations, say measurements affected by errors, can be presented as a *histogram* provided a sufficiently large number of observations has been recorded.

Denote in general the observations by X_k and define intervals ΔX_k centred around X_k, where subscript $k = 1, 2, \ldots, N$ with $N =$ total number of intervals. Now count the number of observations that fall into every interval ΔX_k and call it $m_k = $*frequency of occurrence* per interval. Plotting m_k for all intervals ΔX_k across the entire range of observations X_k and presenting the diagram in the form of a collection of vertical bars, as shown in Fig. A2.1, constitutes the histogram. A line plot joining consecutively all top centres of the histogram bars is known as the *frequency distribution* plot which can also be seen in Fig. A2.1. The latter plot, as will be shown in

Fig. A2.1. Frequency distribution (histogram).

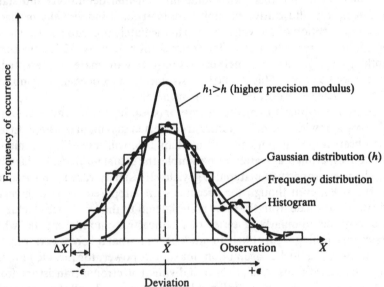

Section A2.4, is relevant when experimental frequency distributions are compared with theoretical (idealised) models.

A2.3 The mean and the measures of deviation

One natural tendency when appraising scattered data is to calculate first the respective *arithmetic mean*, hereafter called simply the mean. Incidentally, it can be shown by rigorous statistical speculations to be the *most probable* value, so the mean appears to represent a very important statistical parameter. Before the well-known formula for the mean is once again written out here, one should be able to distinguish two possible types of 'collections of items or events' for which the mean may be required. Firstly, one may be faced with an almost infinitely large *population* whose size has but negligible effect upon the mean. Such a mean will be denoted by μ. It is a sort of absolute mean. Secondly, there is a sample mean

$$\bar{X} = \frac{X_1 + X_2 + \cdots + X_n}{n} = \frac{1}{n} \sum_1^n X_k, \tag{A2.1}$$

which refers to a finite set of n observations $X_1, X_2, \ldots, X_n$. The size n does affect $\bar{X}$. Obviously for very large samples $\bar{X} \to \mu$.

Using the concept of frequency of occurrence,

$$\bar{X} = \frac{m_1 X_1 + m_2 X_2 + \cdots + m_N X_N}{m_1 + m_2 + \cdots + m_N}, \tag{A2.1a}$$

where $n = m_1 + m_2 + \cdots + m_N$.

In order to quantify a measure of deviation from the mean for the same sample of n observations one wishes to define the basic deviation

$$\varepsilon_k = X_k - \bar{X}. \tag{A2.2}$$

It would be futile to try to calculate the arithmetic mean of the basic deviations as it can be shown on sight that $\bar{\varepsilon}$ equals identically zero. In electronics one frequently meets with alternating signals and it was for this purpose that the concept of the *root-mean-square* (RMS) value was invented. To obtain the RMS deviation, one squares all deviations ε_k, sums them together, divides the resultant sum by the total number of observations, and finally takes the square root of the result in order to return to the correct order of magnitude.

$$D = \left(\frac{\varepsilon_1^2 + \varepsilon_2^2 + \cdots + \varepsilon_n^2}{n} \right)^{1/2} = \left[\frac{1}{n} \sum_1^n \varepsilon_k^2 \right]^{1/2}. \tag{A2.3}$$

The quantity D is called the *RMS deviation* of a sample.

Statisticians argue that it is not strictly correct to use the total number of observations n for the purposes of finding the measure of deviation for a sample of observations possesses n *degrees of freedom*, one of which is removed by calculating $\bar{X}$. One should, they say, average the sum of squared deviations over the number of remaining degrees of freedom, i.e. by $n-1$. Therefore the accepted measure of deviation of a finite sample is:

$$S = \left[\frac{1}{n-1}\sum_1^n \varepsilon_k^2\right]^{1/2} = \left(\frac{n}{n-1}\right)^{1/2} D, \tag{A2.4a}$$

which is referred to as the sample *standard deviation* and its square as the sample *variance*. The factor $n/(n-1)$ is called *Bessel's correction*.

A similar quantity evaluated for a population would be the population standard deviation σ representing the measure of deviation from μ.

Both sample mean $\bar{X}$ and sample standard deviation S vary from sample to sample, but they tend to μ and σ for $n \to \infty$. Also the difference between the RMS deviation D and S becomes very small for large samples: $(S-D)/S = 0.025$ for $n = 20$ and it tends rapidly to zero with $n \to \infty$.

There is an alternative measure of deviations from the mean, namely the *mean absolute deviation* (MAD)

$$|\varepsilon| = \frac{1}{n}\sum_1^n |\varepsilon_k|, \tag{A2.4b}$$

which for obvious reasons is never zero save for the trivial case of all $\varepsilon_k = 0$.

One interesting feature of the mean $\bar{X}$ is that the sum of squared deviation $\sum \varepsilon_k^2$ can be shown to be always minimum (see also Section A2.10) which is another indication that $\bar{X}$ must indeed be the most probable value.

A2.4 The Gaussian law of errors as a theoretical model for assessing observations

Seemingly it is a law of nature that almost all unbiased observations assume their frequency distribution plots similar to that portrayed previously in Fig. A2.1. There is a well-known analytical curve, known as the Gaussian curve, whose symmetric bell-shaped appearance renders it particularly suitable as a 'templet' for judging the experimental frequency distributions. Changing in Fig. A2.1 the abscissa coordinates from X to ε such that the origin of εs coincides with $\bar{X}$, the frequency of occurrence (signified by y) becomes a function of ε.

Apply now the Gaussian curve to a population in order to get the desired theoretical model for the frequency distribution of deviations $\varepsilon = X - \mu$,

$$y(\varepsilon) = \frac{h}{\sqrt{\pi}} e^{-h^2\varepsilon^2}. \tag{A2.5}$$

Symbol h is here a parameter representing the geometric 'spread' of the Gaussian curve. A formal derivation of equation (A2.5), as a frequency distribution model, is possible using the conventional probability argument. It can be found, for example, in Beers (Ch. 8, Ref. [2]).

There is obviously an inconsistency in applying a continuous function (equation (A2.5)) as a model to a discrete collection of observations. The inconsistency is, however, of secondary importance as long as the number of observations is sufficiently large.

Adopting the Gaussian model without further reservations, one specifies now a deviation interval $\Delta\varepsilon$ centred around some deviation ε. The ratio of the number of deviations that can be found within $\Delta\varepsilon$ to the total number of deviations involved, n,

$$P(\varepsilon) = \frac{m}{n},$$

is the *probability* that an observation X will differ from its mean $\bar{X}$ or μ (whichever case it may be) by $\varepsilon \pm \Delta\varepsilon/2$. The probability $P(\varepsilon)$ can be hence identified with the non-dimensionalised frequency distribution $y(\varepsilon)$ so that the number of observations whose deviations fall into the interval $\Delta\varepsilon$ is:

$$n y(\varepsilon)\, \Delta\varepsilon = \frac{nh}{\sqrt{\pi}}\, e^{-h^2\varepsilon^2}\, \Delta\varepsilon.$$

The population standard deviation therefore reads:

$$\sigma = \lim_{n \to \infty} \frac{1}{n-1} \int_{-\infty}^{+\infty} \varepsilon^2 \frac{nh}{\sqrt{\pi}}\, e^{-h^2\varepsilon^2}\, \mathrm{d}\varepsilon = \frac{1}{h\sqrt{2}}. \tag{A2.6}$$

The parameter h is responsible, as already mentioned, for the shape of the frequency distribution curve. The higher its value the more peaky the distribution curve which in turn means a more precise set of observations. A high frequency of small deviations indicates a high proportion of data lying close to the mean. Hence the parameter h bears the name *precision modulus*.

Because of its direct application in the theory of errors the Gaussian model appears in the literature under the name of *Gaussian law of errors*. The function $y(\varepsilon)$ is correspondingly called the *Gaussian or normal frequency distribution*.

Other forms of the frequency distributions may emerge sometimes in practice, especially when a set of observations is for some reasons biased in one or more directions. This is why there is a need for different distribution models to cope with such non-Gaussian situations. Here one should mention the binomial, hypergeometric and Poisson distributions. Unlike the normal distribution, these distributions are discrete since they entail various forms of series expansions. Their applications range from the sampling procedures through problems of failure to establishing the probability of accidents in time. The reader is at this point referred to literature, for example Chatfield (Ref. [1]) or Guttman *et al.* (Ref. [2]), for further details. Statistical *significance* tests and tests for *goodness-of-fit* are based on two important continuous distribution models: (i) Student's *t*-distribution and (ii) the χ^2-distribution, respectively. The significance tests receive some further attention later in Section A2.11 but regarding the goodness-of-fit procedures the reader once again is advised to consult literature for particulars (e.g. Refs. [1] and [2]).

A2.5 The probable deviations and rejection of observations

Turning now to practical situations, one can think of a situation where two limits, say a_1 and a_2, are specified so that all 'acceptable' deviations must be contained within $a_1 \leqslant \varepsilon \leqslant a_2$. The present desideratum is the probability of occurrence of such deviations. For a population the probability equals the area under the Gaussian curve between the two limits a_1 and a_2. When a_1 and a_2 become $-\infty$ and $+\infty$, respectively, then the area is *unity*, which means 100% probability since all deviations are acceptable.

A customary non-dimensional deviation

$$Z = \frac{X - \mu}{\sigma} = \frac{\varepsilon}{\sigma} \tag{A2.7}$$

transforms equation (A2.5) into

$$y(Z) = \sigma y(\varepsilon) = \frac{1}{\sqrt{(2\pi)}} e^{-z^2/2}, \tag{A2.8}$$

where h has been substituted by equation (A2.6).

The probability area under $y(Z)$ is

$$A = \frac{1}{\sqrt{(2\pi)}} \int_{\alpha_1}^{\alpha_2} e^{-z^2/2} \, dZ, \tag{A2.9}$$

where $\alpha_1 = a_1/\sigma$ and $\alpha_2 = a_2/\sigma$ are the non-dimensional limits of integration. As is well known, there is no explicit solution to integral (A2.9) and its values are available in a tabulated form. An abridged set of tabulated values is given here as Table A2.1 where $\frac{1}{2}A$ stands for integral (A2.9) evaluated between the limits $\alpha_1 = 0$ and $\alpha_2 = \alpha$. This is naturally sufficient as the Gaussian curve is *symmetric* about the origin.

Using Table A2.1 it is evident that symmetric deviation limits $\alpha = \pm 1$ imply, for a population with normal distribution, that 68.26% of all observations will be acceptable. Conversely, if probability of 68.26% is

Table A2.1.

α	$\frac{1}{2}A$	α	$\frac{1}{2}A$	α	$\frac{1}{2}A$	α	$\frac{1}{2}A$
0	0.0000	0.60	0.2257	1.70	0.4554	2.90	0.49813
0.05	0.01994	0.70	0.2580	1.80	0.4641	3.00	0.49865
0.10	0.03983	0.80	0.2881	1.90	0.4713	3.10	0.49903
0.15	0.05962	0.90	0.3159	2.00	0.4772	3.20	0.49931
0.20	0.07926	1.00	0.3413	2.10	0.4821	3.30	0.49952
0.25	0.09871	1.10	0.3643	2.20	0.4861	3.40	0.49966
0.30	0.1179	1.20	0.3849	2.30	0.4893	3.50	0.49977
0.35	0.1368	1.30	0.4032	2.40	0.49180	3.60	0.49984
0.40	0.1554	1.40	0.4192	2.50	0.49379	3.70	0.49989
0.45	0.1736	1.50	0.4332	2.60	0.49534	3.80	0.49993
0.50	0.1915	1.60	0.4452	2.70	0.49653	3.90	0.49995
				2.80	0.49744	4.00	0.49997

required for observations to occur within symmetric limits centred around the origin $\varepsilon = 0$, then it becomes necessary to reject all those observations whose deviations from mean μ fall outside the range $-\sigma \leqslant \varepsilon \leqslant \sigma$. Similarly, the probability of 95.44% imposes limits $-2\sigma \leqslant \varepsilon \leqslant 2\sigma$.

A rapid *rejection* criterion can be evolved from that. The stringency of such a criterion depends solely on the probability desired which itself depends on the individual experimental circumstances. The limits on deviations for the rejection criterion need not be symmetric, of course.

EXAMPLE (A)

A certain dimension of a mass-produced mechanical part has a nominal value of 100 mm with acceptable tolerances of ± 1 mm. Several large samples have been examined resulting in the population mean $\mu = 100.3$ mm and the population standard deviation $\sigma = 0.5$ mm. Calculate the percentage of production that is likely to be rejected.

Limits:

$$a_1 = \frac{(100 - 1) - 100.3}{0.5} = -2.6,$$

$$a_2 = \frac{(100 + 1) - 100.3}{0.5} = 1.4.$$

Using Table A2.1, and recalling the symmetry of the normal distribution, the rejected percentage is

$$100[1 - (0.4192 + 0.4954)] = 8.5\%.$$

A set of observations may sometimes contain an odd value which seems inconsistent with the rest of the data. Often there is a temptation to discard the suspect result. How does one decide whether or not to reject it? This may be an important question, especially for small samples where the probability of occurrence of large deviations is small and retention of such incoherent observations may distort the overall result or conclusion. The answer to this problem is not always easy as clear-cut reasons for taking the correct decisions are rarely at hand.

First of all, an odd-looking observation may be in fact quite legitimate and it is always prudent to check this point. Secondly, a decision is needed as to what probability of occurrence is acceptable or desirable. The choice is rather individual, but unless there are other special reasons one may be justified in rejecting those observations whose deviations are less than 50% probable to happen. This particular rule is fairly popular amongst experimental physicists; example (b) demonstrates its application.

EXAMPLE (B)

An experimental sample consisting of 13 observations of the relative humidity index in some stable environment resulted in the following readings: 49, 51, 50, 47, 52, 57, 53, 48, 50, 52, 54, 51 and 49. At first sight the readings 47 and 57 are potentially suspect. Assuming that no special reasons exist for the deviations, apply the 50% probability criterion to see if any of the two readings could be left out.

Mean:

$$\bar{X} = \tfrac{663}{13} = 51.$$

Standard deviation:

$$S = \sqrt{\tfrac{86}{12}} = 2.68.$$

50% probability criterion:

$$n(1 - A) = \tfrac{1}{2},$$

$$A = \frac{2n - 1}{2n} = \frac{25}{26} = 0.9615,$$

for which value interpolation of Table A2.1 gives $\alpha = \pm 2.07$. Hence the maximum acceptable deviations are

$$\varepsilon_{max} = \pm 2.07 \times 2.68 = \pm 5.55.$$

The reading 57, whose deviation from mean is $57-51 = 6 > 5.55$, can be rejected but the reading 47 giving deviation from mean $47-51 = -4 > -5.55$ should be retained.

A2.6 Checking the applicability of the Gaussian model

Before applying the Gaussian model to the experimental distribution under investigation one must first establish that there is a satisfactory resemblance between the two distributions. Some experimental distributions can be inherently biased and the Gaussian model would be irrelevant. They may be skewed, excessively peaky (leptokurtic) or excessively flat (platykurtic) as presented in Fig. A2.2. Apart from superimposing the distribution and the model on one another graphically and exercising a rather subjective judgement about their mutual resemblance, one can perform a simple statistical test based on two typical 'shape' factors: *the skewness factor λ* and the *flatness factor κ*, which for the normal distribution are known to be 0 and 3 respectively. Both λ and κ are related to the RMS deviation D (cf. equation (A2.3)).

$$\lambda = \frac{1}{nD^3} \sum_1^n \varepsilon_k^3; \quad \kappa = \frac{1}{nD^4} \sum_1^n \varepsilon_k^4. \tag{A2.10}$$

It is probably in order to accept as near enough normal all those experimental distributions for which $-0.3 < \lambda < 0.3$ and $2.5 < \kappa < 3.5$. Different distribution models should be sought, as suggested in Section A2.4, if the above limits are exceeded.

A goodness-of-fit test was also mentioned in Section A2.4. The essence of this test rests on evaluating the parameter

$$\chi^2 = \sum_1^n \frac{(\text{observations} - \text{expected values})^2}{\text{expected values}},$$

where the expected values could be, for example, the postulated Gaussian model. The χ^2-test should be regarded as an alternative or complementary

Fig. A2.2. Abnormal distributions: (*a*) positive skew; (*b*) negative skew; (*c*) peaky; (*d*) flat; (*e*) normal.

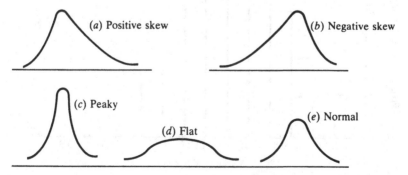

test for checking the applicability of the Gaussian model. The computed χ^2 is normally compared with its value resulting from the χ^2-distribution for an appropriate probability and corresponding number of degrees of freedom (depending on the size of the sample). For working details of the χ^2-test consult the recommended literature.

A2.7   An exercise involving material presented in Sections A2.2–2.6

It may be useful to illustrate the application of the material presented thus far on a practical example.

Suppose that it was required to conduct a hydraulic experiment during which a certain pressure should have been maintained constant and equal to 10 bar (1 bar $= 10^5$ N m^{-2}). However, this condition proved impossible to secure and, as seen in Table A2.2, where results of $n = 50$ consecutive readings are tabulated, some appreciable pressure fluctuations were observed. The experimenter feels that in his report he must document this particular source of error as it affects his further results. He therefore carried out the following calculations:

(1)   *Data presentation:* see Table A2.2 where $m =$ number of occurrences per interval $\Delta X = 0.1$ bar; Fig. A2.3 contains the respective histogram;

Fig. A2.3. Histogram.

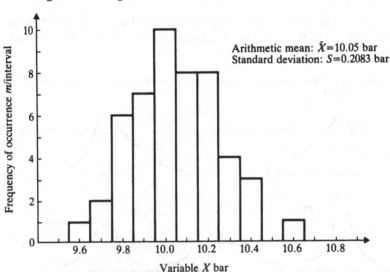

Arithmetic mean: $\bar{X} = 10.05$ bar
Standard deviation: $S = 0.2083$ bar

Frequency of occurrence m/interval

Variable X bar

Table A2.2.

X	m	$\sum m$	mX	$\varepsilon = \bar{X} - X$	ε^2	ε^3	ε^4	$m|\varepsilon|$	$m\varepsilon^2$	$m\varepsilon^3$	$m\varepsilon^4$
9.6	1	1	9.6	−0.45	0.2025	−0.0911	0.0410	0.45	0.2025	−0.0911	0.0410
9.7	2	3	19.4	−0.35	0.1225	−0.0429	0.0150	0.70	0.2450	−0.0858	0.0300
9.8	6	9	58.8	−0.25	0.0625	−0.0156	0.0039	1.50	0.3750	−0.0936	0.0234
9.9	7	16	69.3	−0.15	0.0225	−0.0034	0.0005	1.05	0.1575	−0.0238	0.0035
10.0	10	26	100.0	−0.05	0.0025	−0.0001	0.0000	0.50	0.0250	−0.0010	0.0000
10.1	8	34	80.8	0.05	0.0025	0.0001	0.0000	0.40	0.0200	0.0008	0.0000
10.2	8	42	81.6	0.15	0.0225	0.0034	0.0005	1.20	0.1800	0.0272	0.0040
10.3	4	46	41.2	0.25	0.0625	0.0156	0.0039	1.00	0.2500	0.0624	0.0156
10.4	3	49	31.2	0.35	0.1225	0.0429	0.0150	1.05	0.3675	0.1287	0.0450
10.5	0	49	0	0.45	0.2025	0.0911	0.0410	0	0	0	0
10.6	1	50	10.6	0.55	0.3025	0.1664	0.0915	0.55	0.3025	0.1664	0.0915
$\sum$	50	−	502.5	−	−	−	−	8.40	2.1250	0.0902	0.2540

(2)   *Mean* (equation (A2.1a)): $\bar{X} = 502.5/50 = 10.05$ bar;
(3)   *RMS deviation* (equation (A2.3)): $D = \sqrt{(2.125/50)} = 0.2062$ bar;
(4)   *MAD* (equation (A2.4)): $|\varepsilon| = 8.40/50 = 0.168$ bar;
(5)   *Standard deviation* (equation (A2.4a)): $\sigma = \sqrt{(50/49)}0.2062 = 0.2083$ bar;
(6)   *Precision modulus* (equation A2.6)): $h = 1/0.2083\sqrt{2} = 3.39$ bar^{-1};
(7)   *Gaussian curve:* probability per interval $\Delta\varepsilon = 0.1$,

$$y\,\Delta\varepsilon = \frac{3.39 \times 0.1}{\sqrt{\pi}}\,e^{-3.39^2\varepsilon^2} = 0.191\,e^{-11.5\varepsilon^2}.$$

See Fig. A2.4.
(8)   *Skewness and flatness factors* (equation (A2.10)):

$$\lambda = \frac{0.0902}{50 \times 0.2062^3} = 0.206; \quad \kappa = \frac{0.2540}{50 \times 0.2062^4} = 2.81.$$

Neither λ nor κ are ideally Gaussian (0 and 3, respectively) but within the limits recommended in Section A2.6.

A2.8   The standard error

If random samples, each of size n, are drawn from a population whose mean is μ and standard deviation is σ, then the sample means $\bar{X}_i = \mu$ and their

Fig. A2.4. Gaussian model fit.

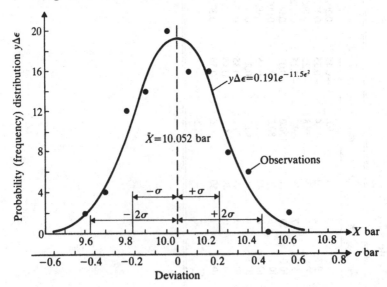

standard deviation, known as the *standard error*, is

$$\sigma_m \simeq \frac{\sigma}{\sqrt{n}}. \tag{A2.11}$$

Further, if similar random samples are drawn from a population with a normal distribution, then the distribution of means $\bar{X}_i$ is also normal with an identical standard error σ_m.

Equation (A2.11) is only approximate but its accuracy improves with the sample size n. It is considered to be correct for $n > 20$.

EXAMPLE (C)

In order to establish a routine checking of the concentration of a certain chemical in a substance manufactured on a production line, one takes a series of measurements at regular intervals. Knowing from previous samples that there is no systematic error involved and that the standard deviation per sample is 2.5%, calculate what should be the minimum size of the regular test samples so that the standard error of the concentration be kept under 0.5%.

Using equation (A2.11),

$$n \geqslant \left(\frac{\sigma}{\sigma_m}\right)^2 = \left(\frac{2.5}{0.5}\right)^2 = 25.$$

Minimum sample size is 25.

A2.9 Propagation of errors

To determine experimentally the power of a gasoline motor it is necessary to measure the torque on, and the rotational speed of, the driving shaft (cf. Appendix 1). This is a typical example where an experimental quantity is not measured directly, but calculated from a formula using several constituent measurements.

$$Q = Q(X, Y, Z, \ldots).$$

All these constituent measurements are subject to their individual errors, $\Delta X, \Delta Y, \Delta Z, \ldots$, thus inflicting an error ΔQ upon the quantity Q. The objective of this section is to show how ΔQ can be estimated.

Apply in this case the well-known calculus formula for a small increment of a function of many variables

$$\Delta Q = \frac{\partial Q}{\partial X}\,\Delta X + \frac{\partial Q}{\partial Y}\,\Delta Y + \frac{\partial Q}{\partial Z}\,\Delta Z + \cdots, \tag{A2.12}$$

where the partial derivatives, $\partial Q/\partial X$, $\partial Q/\partial Y$, ..., must be evaluated within the intervals, ΔX, ΔY, ... The following Example (d) explains how the procedure works.

EXAMPLE (D)

The classical simple pendulum experiment may be used to determine the gravitational acceleration g. Assuming that it is possible to measure the length of the pendulum with virtually no error at all, find what maximum error can be allowed for the time measurement in order to secure an accuracy of 1% for g.

The pendulum equation is:

$$T = \pi \sqrt{\left(\frac{L}{g}\right)},$$

where L = pendulum length; T = period of oscillations. Hence

$$g = \pi^2 \frac{L}{T^2},$$

$$\Delta g = \pi^2 \left(\frac{\Delta L}{T^2} - 2 \frac{L}{T^3} \Delta T\right),$$

$$\frac{\Delta g}{g} = \frac{\Delta L}{L} - 2 \frac{\Delta T}{T} \simeq -2 \frac{\Delta T}{T},$$

since $(\Delta L)/L \ll (\Delta T)/T$ by assumption. In order to attain an accuracy of 1% for g, the period T must be measured with an accuracy of at least $\frac{1}{2}$%.

In the above example it has been assumed that the error tolerances ΔL and ΔT are well defined. This condition could be difficult to satisfy in practice. How, for example, can one guarantee that certain error tolerances will never be surpassed? It is often safer, therefore, to replace the absolute tolerances in equation (A2.12) with the respective standard error

$$\sigma_{m_Q}^2 = \left(\frac{\partial Q}{\partial X} \sigma_{m_X}\right)^2 + \left(\frac{\partial Q}{\partial Y} \sigma_{m_Y}\right)^2 + \cdots, \tag{A2.13}$$

which mathematical transformation can be strictly justified.

The following rules apply to the four basic algebraic operations on the constituent measurements:

(1)   *Addition and/or subtraction:*

$$Q = a \pm b \pm c \pm \cdots,$$

$$\sigma_{m_Q}^2 = \sigma_{m_a}^2 \pm \sigma_{m_b}^2 \pm \sigma_{m_c}^2 \pm \cdots.$$

(2) *Multiplication and/or division of powers* $(\alpha, \beta, \gamma, \ldots)$:

$$Q = a^\alpha b^\beta c^\gamma \cdots.$$

$$\sigma_{m_Q}^2 = (a^\alpha b^\beta c^\gamma \cdots)^2 \left[\left(\frac{\alpha}{a} \sigma_{m_a} \right)^2 + \left(\frac{\beta}{b} \sigma_{m_b} \right)^2 + \left(\frac{\gamma}{c} \sigma_{m_c} \right)^2 + \cdots \right].$$

EXAMPLE (E)

Derive an expression for the standard error of the quantity Q

$$Q = A^2 B^{-1/2}$$

in terms of the constituent σ_{m_A} and σ_{m_B}:

$$\sigma_{m_Q}^2 = A^4 B^{-1} \left(\frac{4}{A^2} \sigma_{m_A}^2 + \frac{1}{4B} \sigma_{m_B}^2 \right) = \frac{4A^2}{B} \sigma_{m_A}^2 + \frac{A^4}{4B^2} \sigma_{m_B}^2.$$

A2.10 Curve fitting by least squares

Curve fitting by eye is probably by far the most common method of fixing an empirical relationship for two sets of mutually dependent variables. For high precision experiments where the scatter of points is not too great, the procedure is in fact quite satisfactory. It may not be, however, for low precision experiments yielding widely scattered results. The process of estimating the functional dependence of an empirical relationship using a set of values for a sample is known as the *regression analysis*.

Probably the most common regression analysis is the well-established *least squares* procedure where the key feature is that the sum of all deviations between the actual coordinates of the dependent variables and the coordinates of the regression curve (often called the *residuals*) *must be minimum*.

Consider a set of n experimental points, for example as in Fig. A2.5. The respective residuals are

$$\delta_k = f(X_k) - Y_k \quad (k = 1, 2, \ldots, n),$$

where $(X_1, Y_1), (X_2, Y_2), \ldots, (X_n, Y_n)$ are the experimental points and $Y = f(X)$ is the regression function to be fitted. The function is a least squares regression if

$$\delta_1^2 + \delta_2^2 + \cdots + \delta_n^2 = \text{minimum}.$$

Polynomial functions $f(X)$ are particularly suitable for this purpose:

$$Y = C_0 + C_1 X + C_2 X^2 + \cdots + C_i X^i = \sum_0^i C_k X^k. \tag{A2.14}$$

Here i denotes the order of the polynomial. The $i+1$ coefficients C_0 through C_i are calculated from the condition of minimum residuals,

$$R = \sum_0^i \delta_k^2 = \sum_0^i [f(X_k) - Y_k]^2 = \text{minimum},$$

which is satisfied by

$$\frac{\partial R}{\partial C_0} = \frac{\partial R}{\partial C_1} = \cdots = \frac{\partial R}{\partial C_i} = 0. \tag{A2.15}$$

The necessary requirement obviously is that the number of experimental points available for the polynomial regression analysis is at least one more than the order of the polynomial to be fitted, i.e. $n \geqslant i+1$.

Conditions (A2.15) yield the following expressions for the coefficients of an ith-order polynomial (equation (A2.14)).

$$C_0 n + C_1 \sum_1^n X_k + \cdots + C_i \sum_1^n X_k^i = \sum_1^n Y_k$$

$$C_0 \sum_1^n X_k + C_1 \sum_1^n X_k^2 + \cdots + C_i \sum_1^n X_k^{i+1} = \sum_1^n X_k Y_k$$

$$\vdots \tag{A2.16}$$

$$C_0 \sum_1^n X_k^i + C_1 \sum_1^n X_k^{i+1} + \cdots + C_i \sum_1^n X_k^{2i} = \sum_1^n X_k^i Y_k$$

which for a straight line ($Y = C_0 + C_1 X$) degenerate to

$$C_0 n + C_1 \sum_1^n X_k = \sum_1^n Y_k$$
$$C_0 \sum_1^n X_k + C_1 \sum_1^n X_k^2 = \sum_1^n X_k Y_k \tag{A2.17}$$

The higher the order of the polynomial the more laborious are the calculations leading to the solution of a set of $i+1$ algebraic equations (A2.16). There are special numerical subroutines to do the job on a computer, but it would be foolish to try always to fit the highest-order polynomial possible ($i = n - 1$). A quick visual examination of the distribution of the experimental points considered, often suggests a low-order polynomial (straight line, parabola, cubic) which could be quite adequate. One should realise that higher-order polynomials fitted by the least squares technique frequently tend to undulate, which feature may be quite wrong for the physical phenomenon that the particular regression is meant to represent.

When fitting the regression curves one must always decide early which of the interdependent variables is going to be the independent variable. The choice is not always obvious, and it seems best to call the independent variable that one which is easier to measure.

The effect of the condition of minimum residuals in least squares procedures is that the resultant Ys are statistically most probable. Fig. A2.6

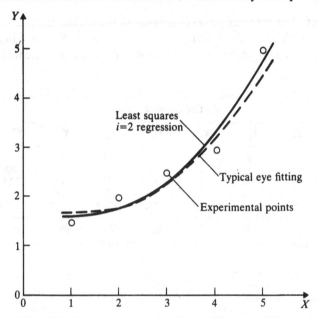

Fig. A2.5. Least squares curve fitting [Example (f)].

Fig. A2.6. Distribution of Y at X = constant.

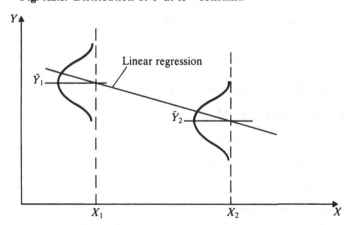

shows this feature by indicating the respective frequency distributions which concept is very pertinent here. In Section A2.12 the problem of confidence intervals for the linear regression will be discussed further.

EXAMPLE (F)

A geometric contour is defined by a set of approximate coordinates (X_k, Y_k), as shown in Fig. A2.5. They are $(1, 1.5), (2, 2), (3, 2.5), (4, 3), (5, 5)$. Try to fit a least squares parabola to these points. The procedure is given in the table below.

		Coefficients C_0, C_1, C_2				
X_k	Y_k	X_k^2	X_k^3	X_k^4	$X_k Y_k$	$X_k^2 Y_k$
1	1.5	1	1	1	1.5	1.5
2	2.0	4	8	16	4.0	8.0
3	2.5	9	27	81	7.5	22.5
4	3.0	16	64	256	12.0	48.0
5	5.0	25	125	625	25.0	125.0
$\sum$ 15	14	55	225	979	50	205

	Regression		
C_0	$C_1 X$	$C_2 X^2$	$'Y$
1.9	0.4857	0.2143	1.629
1.9	0.9714	0.8572	1.786
1.9	1.4571	1.9286	2.371
1.9	1.9428	3.4386	3.386
1.9	2.4286	5.3572	4.829

The equations for the coefficients C_0, C_1, C_2

$$C_0 5 + C_1 15 + C_2 55 = 14,$$

$$C_0 15 + C_1 55 + C_2 225 = 50,$$

$$C_0 55 + C_1 225 + C_2 979 = 205,$$

have a solution:

$$C_0 = 1.9; \quad C_1 = -0.4857; \quad C_2 = 0.2143.$$

Hence, the least squares parabola is

$$Y = 1.9 - 0.4857\, X + 0.2143\, X^2.$$

The results are presented in Fig. A2.5 where an intuitive eye fit is also superimposed for comparison.

A2.11 The significance tests

Means of detecting whether differences between two sets of observations are due to a real reason or merely due to a chance factor are often vital in experimentation. Also two interim inspection samples from the same population are unlikely to be identical and one needs to establish whether the differences are insignificant and incidental or significant indicating progressive permanent changes in the population.

The differences between samples are said to be significant if the probability of getting one as large as, or larger than, the observed ones is less than 5%, i.e. 1 in 20. The 5% probability criterion is not a rigid rule, but a reasonable practical compromise. It involves the risk of a true hypothesis being discarded once in 20 samples.

For a sample of n observations drawn from a Gaussian population the non-dimensional deviation of sample mean $\bar{X}$ from population mean μ is, according to equations (A2.7) and (A2.11),

$$Z = \sqrt{n}\,\frac{\bar{X} - \mu}{\sigma}.$$

It is not common to know the population standard deviation σ so a similar non-dimensional deviation based on the sample standard deviation,

$$t = \sqrt{n}\,\frac{\bar{X} - \mu}{S}, \tag{A2.18}$$

is better used.

The frequency distribution of the deviation, t, is known to be, for small size samples ($n < 20$), distinctly different from the normal distribution. The corresponding theoretical model, shown in Fig. A2.7, is not as peaky as the Gaussian model and it depends strongly on the number of degrees of freedom v of the sample concerned (cf. Section A2.3). For large vs, i.e. large samples, the t-distribution becomes asymptotically Gaussian.

Here again the area under the t-curve represents the probability of a deviation to occur. The symbol α represents conventionally the combined areas under the t-curve from $-\infty$ to $-t_v$ and from t_v to $+\infty$ (see Fig. A2.7) and as such is called the *significance level* α. In line with the above probability criterion, the parameter α indicates the probability of getting the sample mean deviation $\bar{X} - \mu$ equal to or greater than

$$\frac{t_{\alpha, v} s}{\sqrt{n}}.$$

Fig. A2.7. *t*-Significance test: values for $t_{\alpha, v}$ [$\alpha = 0.001$ (very significant); $\alpha = 0.01$ (significant); $\alpha = 0.05$ (rather significant)].

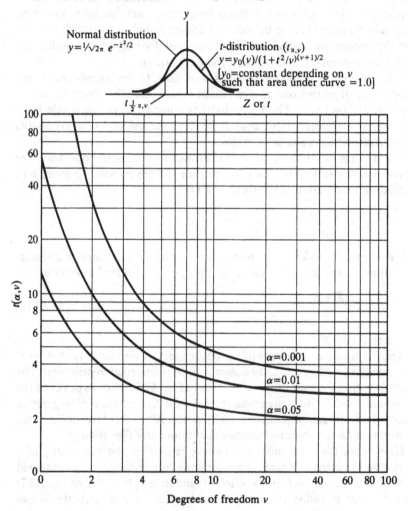

Alternatively, $1-\alpha$ signifies the *confidence* that the sample mean deviation $\bar{X} - \mu$ will not exceed the limits

$$\pm \frac{t_\alpha, v^s}{\sqrt{n}}.$$

This is in fact nothing more than an extension of the procedure evolved in Section A2.5 for normal distributions. Presently it is being applied to small-size samples under the name of *t-significance tests*. The corresponding nomogram for three significance levels $\alpha = 0.05$, 0.01 and 0.001 is given in Fig. A2.7.

EXAMPLE (G)

The lift coefficient on an aerofoil has been calculated using two independent theoretical methods A and B. For a given incidence angle, theory A predicts $C_L = 1.81$ and theory B predicts $C_L = 1.87$. To verify both theories a model of the aerofoil has been tested at the same incidence angle and the corresponding flow conditions (Reynolds number, etc.). The following five observations have been recorded for C_L: 1.75, 1.80, 1.82, 1.72 and 1.76. Apply the *t*-significance test to establish if any of the theoretical results are significantly different from the observations.

(1) *Mean:* $\bar{X} = \dfrac{1.75 + 1.80 + 1.82 + 1.72 + 1.76}{5} = \dfrac{8.85}{5} = 1.77.$

(2) *Standard deviation:* $\sum \varepsilon^2 = 0.0064$

$$S = \sqrt{\left(\frac{0.0064}{5-1}\right)} = 0.04.$$

(3) *t-test* (equation (A2.18)):

$$t_A = \frac{1.77 - 1.81}{0.04}\sqrt{5} = -2.24 \quad \text{(method } A\text{)}$$

$$t_B = \frac{1.77 - 1.87}{0.04}\sqrt{5} = -5.59 \quad \text{(method } B\text{).}$$

From the curves in Fig. A2.7 the 5% and 1% probabilities of observing deviations equal to or larger than t for a sample with four degrees of freedom give:

$$t_{0.05, 4} = 2.78 \quad \text{and} \quad t_{0.01, 4} = 4.60.$$

Clearly, method A yields a result that is not significantly different from the observations as $|t_A| < t_{0.05, 4}$ and $< t_{0.01, 4}$. Method B, however, disagrees significantly with the observations as $|t_B| > t_{0.05, 4}$ and $> t_{0.01, 4}$.

The systematic experiment

EXAMPLE (H)

A sample of ten measurements of length of a certain mechanical part resulted in mean $\bar{X} = 5.3$ mm and standard deviation $S = 0.08$ mm. Find the 95% and 99% confidence limits for the actual length, i.e.

$$\bar{X} \pm t_{\alpha,v} S/\sqrt{n}.$$

Here the number of degrees of freedom $v = n - 1 = 9$.

95% confidence limits: $\alpha = 0.05$;

$t_{0.05,\,9} = 2.26$,

$$5.31 \pm \frac{2.26 \times 0.08}{\sqrt{9}} = 5.31 \pm 0.06.$$

99% confidence limits: $\alpha = 0.01$;

$t_{0.01,\,9} = 3.25$,

$$5.31 \pm \frac{3.25 \times 0.08}{\sqrt{9}} = 5.31 \pm 0.09.$$

A2.12 The confidence limits in linear regression

Further to section A2.10, the constants C_0 and C_1 for a least squares line according to equation (A2.17) are:

$$C_0 = \frac{\sum Y_k \sum X_k^2 - \sum X_k \sum X_k Y_k}{n \sum X_k^2 - (\sum X_k)^2},$$

$$C_1 = \frac{n \sum X_k Y_k - \sum X_k \sum Y_k}{n \sum X_k^2 - (\sum X_k)^2}.$$

In order to establish the confidence limits (boundaries), say at the level $100(1-\alpha)\%$, for the linear regression one calculates first the standard deviation of the residuals

$$S_y^2 = \frac{1}{n-2} \sum_1^n [Y_k - (C_0 + C_1 X_k)]^2, \tag{A2.19}$$

where $n - 2$ appears because two degrees of freedom have already been used for calculating C_0 and C_1.

Now, since the total number of points used for the purpose of regression may be typical of a small sample, i.e. less than 20, it is proper to use the t-distribution rather than the normal distribution for evaluating the confidence boundaries. Their equation is expressed here in terms of the variables $\hat{Y} = f(\hat{X})$. The confidence boundaries are

$$\hat{Y} = C_0 + C_1 \hat{X} \pm t_{\alpha, n-2} S_y \left[\frac{1}{n} + \frac{(\hat{X} - \bar{X})^2}{\sum_1^n (X_k - \bar{X})^2} \right]^{1/2}. \tag{A2.20}$$

Equation (A2.20) is quoted without derivation; it defines two separate boundaries: one below and one above the linear regression $Y = C_0 + C_1 X$. Note that the confidence limits are smallest at the point where $\hat{X} = \bar{X}$. A typical formation of such confidence boundaries is illustrated in Fig. A2.8.

Similarly, the confidence boundaries can be evaluated for higher-order regressions as well. However, the problem becomes then algebraically quite complex and tedious and numerical procedures based on computer facilities are best advised in those circumstances. For particulars consult more advanced textbooks (e.g. Guttman *et al.*, Ref. [2]).

A2.13 Closing remarks

For obvious reasons, this appendix fails to deal with many other important aspects of error analysis and random data evaluation. No, or but casual, mention is made about such problems as: (i) correlated or interdependent

Fig. A2.8. Confidence limits for linear regression using $t_{\alpha, v}$ *significance values.*

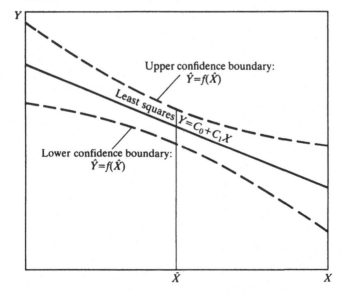

observations and errors; (ii) weighted measurements where some observations are trusted more than the others and therefore are given higher weighting factors; (iii) goodness-of-fit tests. Neither does this appendix explore the wide scope of other useful distributions (Poisson, binomial, etc.), nor does it emphasise strongly enough the role of the probability theory in processing and interpreting experimental observations. The reader is hereby encouraged to further his horizons in this area by studying the subject matter beyond the depth of the present exposition. A thorough ability of analysing and interpreting experiments and their inherent errors should be recognised as a worthwhile personal investment.

References

[1]   Chatfield, C. (1970). *Statistics for Technology.* Penguin Books.
[2]   Guttman, I. Wilks, S. S. & Hunter, J. S. (1971). *Introductory Engineering Statistics.* Wiley.

Appendix 3
Notation, units, laws, formulae, properties and constants

Although Imperial and other units will be encountered in laboratory work, SI units will generally be used.

The eight primary units are listed in Table A3.1, together with some special derived units. Fractions and multiples, and some conversion factors are also listed.

Table A3.1. *SI units and Imperial units*

SI units: the eight primary units

Quantity	Unit	Symbol
length	metre	m
mass	kilogram	kg
time	second	s
electric current	ampere	A
thermodynamic temperature	kelvin	K
luminous intensity	candela	cd
amount of substance	mole	mol
plane angle	radian	rad

Some special derived units

Quantity	Unit	Symbol	Definition
solid angle	steradian	sr	$m^2/4\pi\ m^2$
frequency	hertz	Hz	s^{-1}
force	newton	N	$kg\ m\ s^{-2}$
pressure, stress	pascal	Pa	$N\ m^{-2}$
dynamic viscosity	pascal second	Pa s	$Ns\ m^{-2}$

Table A3.1 (*cont.*)

work, energy, quantity of heat	joule	J	Nm
power, heat flow rate	watt	W	J s^{-1}
electric charge	coulomb	C	As
electric potential difference	volt	V	W A^{-1}
electric resistance	ohm	Ω	V A^{-1}
electric conductance	siemens	S	A V^{-1}
electric capacitance	farad	F	As V^{-1}
magnetic flux	weber	Wb	Vs
inductance	henry	H	Vs a^{-1}
magnetic flux density	tesla	T	Wb m^{-2}
luminous flux	lumen	F	cd sr^{-1}
illuminance	lux	E	F m^{-2}

Fractions and multiples

Fraction	Prefix	Symbol	Multiple	Prefix	Symbol
10^{-1}	deci	d	10	deca	da
10^{-2}	centi	c	10^2	hecto	h
10^{-3}	milli	m	10^3	kilo	k
10^{-6}	micro	μ	10^6	mega	M
10^{-9}	nano	n	10^9	giga	G
10^{-12}	pico	p	10^{12}	tera	T
10^{-15}	femto	f			
10^{-18}	atto	a			

Imperial units: the six primary units are

Quantity	Unit	Symbol
length	foot	ft
mass	pound	lbm
time	second	s
electric current	ampere	A
temperature	degree Rankine	°R
luminous intensity	candela	cd

Some special derived units

Quantity	Unit	Symbol	Definition
force	pound	lbf	$(1/g_0)$ lbm ft s^{-2}
work, energy	foot pound	ft lbf	ft lbf
quantity of heat	British thermal unit	Btu	J ft lbf
power	horsepower	hp	550 ft lbf s^{-1}

Notes:
1. $(1/g_0)$ is the constant or proportionality in Newton's second law of motion so that

$$1 \text{ lbf} = \frac{1 \text{ lbm} \times 32.174 \text{ ft s}^{-2}}{g_0}.$$

Table A3.1 (*cont.*)

Therefore

$$g_0 = 32.174 \frac{\text{lbm ft}}{\text{lbf s}^2}.$$

2. J is Joule's equivalent and relates heat units to work units in the Imperial system so that

$$1 \text{ Btu} = 778 \text{ ft lbf}$$

or

$$J = 778 \text{ ft lbf Btu}^{-1}.$$

Conversion factors (four-figure accuracy)

	Imperial unit		SI unit
length	1 in	=	2.540 cm
	1 ft	=	0.3048 m
	1 mile	=	1.609 km
area	1 in^2	=	6.452 cm^2
	1 ft^2	=	0.092 90 m^2
volume	1 in^3	=	16.39 cm^3
	1 ft^3	=	0.028 32 m^3
velocity	1 ft s^{-1}	=	0.3048 m s^{-1}
	1 mile h^{-1}	=	1.609 km h^{-1}
mass	1 lbm	=	0.4536 kg
	1 slug	=	14.59 kg
density	1 lbm ft^{-3}	=	16.02 kg m^{-3}
	1 slug ft^{-3}	=	515.4 kg m^{-3}
force	1 lbf	=	4.448 N
pressure	1 lbf in^{-2}	=	6.895 kPa
	1 in H$_2$O	=	249.1 Pa
	1 in Hg	=	3.386 kPa
energy	1 ft lbf	=	1.356 J
	1 Btu	=	1.055 kJ
power	1 hp	=	745.7 W

Although notation will vary from person to person, the notation and associated units given in Table A3.2 are widely used.

Several fundamental laws and derived relations, which will prove useful during experimental work, are summarised on subsequent pages. Finally, some standard values, properties and constants are quoted.

Table A3.2. *List of commonly used symbols*

Symbol	Quantity	Usual units SI	Imperial
A	area	m^2	ft^2
a	sonic velocity	m s^{-1}	ft s^{-1}
B	intensity of magnetic field	T	

Table A3.2 (*cont.*)

Symbol	Quantity	SI	Imperial
		Usual units	
b	breadth	m	ft
C	capacitance	F	
C_p	specific heat at constant pressure	kJ kg^{-1} K^{-1}	Btu lbm^{-1} °R^{-1}
C_v	specific heat at constant volume	kJ kg^{-1} K^{-1}	Btu lbm^{-1} °R^{-1}
d	diameter, depth	m	ft
	distance between plates of condenser	m	
E	energy	kJ	ft lbf, Btu
	modulus of elasticity	GPa	lbf in^{-2}
	intensity of electric field	V m^{-1}	
e	specific energy	kJ kg^{-1}	ft lbf lbm^{-1}
F	force	N	lbf
f	acceleration	m s^{-2}	ft s^{-2}
	frequency	Hz	s^{-1}
G	shear modulus	GPa	lbf in^{-2}
g	gravitational acceleration	m s^{-2}	ft s^{-2}
H	enthalpy	kJ	Btu
	angular momentum	kg m^2 s^{-1}	lbm ft^2 s^{-1}
	height	m	ft
h	specific enthalpy	kJ kg^{-1}	Btu lbm^{-1}
	heat transfer coefficient	W m^{-2} K^{-1}	Btu ft^{-2}h^{-1} °R^{-1}
I	second moment of area	m^4	ft^4
	moment of inertia	kg m^2	lbm ft^2
	electric current	A	
k	radius of gyration	m	ft
	thermal conductivity	W m^{-1} K^{-1}	Btu ft^{-1} h^{-1} °R^{-1}
L	self-inductance	H	
l	length	m	ft
M	Mach number	—	—
	moment	N m	lbf ft
	mutual inductance	H	
$\overline{M}$	molecular mass	—	—
m	mass	kg	lbm
N	angular velocity	rev min^{-1}	rev min^{-1}
n	frequency	Hz	s^{-1}
P	stagnation pressure	kPa	lbf in^{-2}
p	pressure	kPa	lbf in^{-2}
Q	heat transfer	kJ	Btu
	volume flow rate	m^3 s^{-1}	ft^3 s^{-1}
	isolated electric point charge	C	
q	heat transfer per unit mass	kJ kg^{-1}	Btu lbm^{-1}
	electric point charge	C	
R	gas constant	kJ kg^{-1} K^{-1}	Btu lbm^{-1} °R^{-1}
	resistance	Ω	

Table A3.2 (*cont.*)

Symbol	Quantity	SI	Imperial
		Usual units	

Symbol	Quantity	SI	Imperial
R	molal gas constant	kJ kg-mol^{-1} K^{-1}	Btu lb-mol^{-1} °R^{-1}
r	radius	m	ft
	distance between electric charges	m	
S	entropy	kJ K^{-1}	Btu °R^{-1}
	surface area	m^2	ft^2
s	specific entropy	kJ kg^{-1} K^{-1}	Btu lbm^{-1} °R^{-1}
	distance	m	ft
T	torque	Nm	lbf ft
	temperature	K, °C	°R, °F
	kinetic energy	J	ft lbf
t	time	s	s
U	internal energy, strain energy	kJ	ft lbf, Btu
u	specific internal energy	kJ kg^{-1}	Btu lbm^{-1}
V	velocity	m s^{-1}	ft s^{-1}
	total volume	m^3	ft^3
	shear force	N	lbf
	potential difference	V	
v	specific volume	m^3 kg^{-1}	ft^3 lbm^{-1}
W	work, work transfer	J	ft lbf, Btu
W_s	shaft work	J	ft lbf, Btu
x, y, z	Cartesian coordinates	–	–
y	depth	m	ft
z	expansion	m	ft
α	angular acceleration	rad s^{-2}	rad s^{-2}
	angle	rad	rad
	coefficient of linear expansion	m m^{-1} K^{-1}	in in^{-1} °R^{-1}
γ	ratio of specific heats C_p/C_v	–	–
	shear strain	–	–
δ	deflection	m	ft
	elongation	m	ft
ε	linear strain	–	–
	electromotive force	V	
ε_0	permittivity of free space	pF m^{-1}	
ε_r	dielectric constant	–	
η	efficiency	–	–
θ	angle	rad	rad
λ	wavelength	m	ft
μ	dynamic viscosity	Pa s	slug ft^{-1} s^{-1}
μ_0	permeability of free space	Wb A^{-1} m^{-1}	
v	kinematic viscosity	m^2 s^{-1}	ft^2 s^{-1}
	Poisson's ratio	–	–

Table A3.2 (*cont.*)

Symbol	Quantity	Usual units	
		SI	Imperial
ρ	density	kg m^{-3}	slug ft^{-3}
	charge density	C m^{-3}	
	resistivity	Ωm	
σ	direct stress	Pa	lbf in^{-2}
	surface charge density	C m^{-2}	
τ	shear stress	Pa	lbf in^{-2}
ϕ	flux of magnetic field	Wb	
ω	angular velocity	rad s^{-1}	rad s^{-1}
	circular frequency	rad s^{-1}	rad s^{-1}

Superscripts

$\cdot$ first derivative with respect to time d/dt,

$\cdot\cdot$ second derivative with respect to time d^2/dt^2,

$\hat{\ }$ unit vector,

$^-$ molal quantity.

Vector quantities shown in bold type.

With Imperial units 'inch' and 'foot' can obviously be interchanged (using the appropriate factor of 12) and 'pound mass' (lbm) and 'slug' ($=32.174$ lbm) can also be interchanged.

Fundamental laws
Second law of motion
Force = rate of change of linear momentum with time

$$F = \frac{d}{dt} \ (mV). \tag{A3.1}$$

Torque = rate of change of angular momentum with time

$$T = \frac{dH}{dt}. \tag{A3.2}$$

Conservation of mass
Providing relativistic effects are negligible, the change in mass of a control volume (open system) is equal to the net mass of fluid crossing the boundary of the system.   (A3.3)

First law of thermodynamics
For a closed system, which is taken through any thermodynamic cycle, the cyclic integral of the heat transfers is equal to the cyclic integral of the work transfers.

$$\oint dW = \oint dQ. \tag{A3.4}$$

Second law of thermodynamics

(a) *Kelvin–Planck statement.* It is impossible to construct a device which will operate in a cycle and produce no effect other than the raising of a weight (work transfer) and the exchange of heat with a single reservoir.

$$\tag{A3.5a}$$

(b) *Clausius statement.* It is impossible to construct a device which operates in a cycle and produce no effect other than the transfer of heat from a cooler body to a hotter body.

$$\tag{A3.5b}$$

Equation of state for a perfect gas
For some pure substances, known as perfect or ideal gases, the product or pressure and specific volume is proportional to the absolute temperature

$$pv = \frac{\bar{R}}{M} T. \tag{A3.6}$$

Hooke's law for elastic materials
For an elastic material direct stress is proportional to direct strain.

In one dimension $\sigma = E\varepsilon.$ (A3.7)

For an elastic material shear stress is proportional to shear strain.

In one dimension $\tau = G\gamma.$ (A3.8)

Coulomb's law of electrostatic forces
The electrostatic force between two point charges in a vacuum is directly proportional to the product of their magnitudes and inversely proportional to the square of their separation.

$$F = \frac{q_1 q_2}{4\pi\varepsilon_0 r^2}. \tag{A3.9}$$

Ohm's law of resistance
For electrically conducting materials, the rate of flow of electric charge (or current) is directly proportional to the electric field which, in turn, is directly proportional to the applied potential difference

$$\frac{dq}{dt} = I \propto V \quad \text{or} \quad V = IR. \tag{A3.10}$$

(Although this was an experimentally derived law it can in fact be deduced from the directed motion of electrons under the influence of an electric field.)

Biot–Savart law for magnetic fields
For a current element of length δl which carries a current I the magnetic field intensity due to the element is given by

$$\delta B = \frac{\mu_0 I \delta l \sin \theta}{4\pi r^2}, \tag{A3.11}$$

where θ is the angle between the radius vector r and the current element.

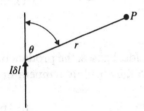

Faraday's law for induced electromotive force
The induced emf is proportional to the time rate of change of the flux of the magnetic field surrounded by the circuit

$$\varepsilon = -\frac{d\phi}{dt}. \tag{A3.12}$$

Definitions
Poisson's ratio for elastic materials
For an elastic material the ratio of lateral strain to longitudinal strain is a constant,

$$v = -\frac{\text{lateral strain}}{\text{longitudinal strain}}. \tag{A3.13}$$

Relation between elastic constants

$$G = \frac{E}{2(1+v)}. \tag{A3.14}$$

Work

The work done by a force is the product of that force and the displacement of the point of action of that force in the direction of the force.

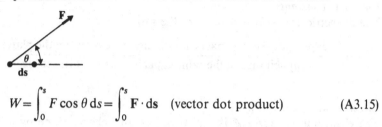

$$W = \int_0^s F \cos \theta \, ds = \int_0^s \mathbf{F} \cdot \mathbf{ds} \quad \text{(vector dot product)} \tag{A3.15}$$

Electric field

The vector force acting on a unit positive charge at any point in space defines the electric field at that point. Thus

$$\mathbf{E} = \frac{\mathbf{F}}{q}. \tag{A3.16}$$

Electrostatic potential

The electrostatic potential (ϕ_p) at a point is the work done per unit charge against the field in bringing a small positive charge from infinity to that point. Thus

$$\phi_p = \frac{W}{q} = -\int_\infty^p \mathbf{E} \cdot \mathbf{dl}. \tag{A3.17}$$

Electron volt

An electron volt is the energy gained by a particle of one electronic charge which is accelerated through one volt potential difference. Thus

$$1 \, \text{eV} = -1.602 \times 10^{-19} \text{ joules.}$$

Capacitance

The capacitance measures the amount of charge necessary to increase the potential of a conductor by one volt. Thus

$$C = \frac{Q}{V}. \tag{A3.18}$$

(Generally μF or pF are most appropriate fractional units.)

Mobility
The mobility (k) is the mean velocity of a charged particle per unit field.
Thus

$$k = \frac{\mathbf{v}}{\mathbf{E}}. \qquad (A3.19)$$

Dielectric constant
The dielectric constant is defined as the ratio

$$\varepsilon_r = \frac{\text{capacitance of a capacitor with dielectric between the plates}}{\text{capacitance of the same capacitor without dielectric}}.$$

Electromotive force (emf)
The electromotive force $\mathscr{E}$ is the work per unit charge done by a non-electrostatic source of energy

$$\mathscr{E} = \frac{dW}{dq}. \qquad (A3.20)$$

Mutual inductance
Mutual inductance (M) is the constant of proportionality which relates the emf induced in one circuit by the rate of change of current in a second circuit

$$\varepsilon_a = -M_{ab}\frac{dI_b}{dt} \quad \text{and} \quad M_{ab} = M_{ba} = M. \qquad (A3.21)$$

Self inductance
Self inductance (L) is the constant of proportionality which relates the emf induced in a single circuit by the rate of change of current in the same circuit.

$$\mathscr{E} = -L\frac{dI}{dt}. \qquad (A3.22)$$

Derived relations
Velocity and acceleration along a straight path

$$f = V = \ddot{s} = \frac{dV}{dt} = \frac{ds}{dt}\frac{dV}{ds} = V\frac{dV}{ds} = \frac{d^2s}{dt^2}, \qquad (A3.23)$$

hence

$$V_2 = V_1 + ft; \quad V_2^2 = V_1^2 + 2fs; \quad \frac{ds}{dt} = V_1 + ft,$$

$$s = V_1 t + \tfrac{1}{2} ft^2, \tag{A3.24}$$

where V_1 is the initial velocity in each case and s is the distance travelled.

Velocity and acceleration along a curved path

$$\mathbf{V} = \dot{r}\hat{\mathbf{r}} + r\dot{\theta}\hat{\boldsymbol{\theta}} \tag{A3.25}$$

$$\mathbf{f} = (\ddot{r} - r\dot{\theta}^2)\hat{\mathbf{r}} + (2\dot{r}\dot{\theta} + r\ddot{\theta})\hat{\boldsymbol{\theta}} \tag{A3.26}$$

centripetal acceleration   coriolis acceleration

Motion of a constant mass
Equation (A3.1) becomes

$$F = m\frac{dV}{dt} = mf = m\ddot{s}. \tag{A3.27}$$

Rotation of a rigid body
Equation (A3.2) becomes

$$T = I\frac{d\omega}{dt} = Ia = I\ddot{\theta} \tag{A3.28}$$

Energy
Kinetic. For a constant mass (A3.27) gives $F = m\ddot{s}$ and for $\theta = 0$ (A3.15) becomes

$$dW = F\,ds = m\ddot{s}\,ds = m\frac{d\dot{s}}{dt}\,ds = m\,d\dot{s}\,\frac{ds}{dt}$$

$$dW = m\dot{s}\,d\dot{s} = d(\tfrac{1}{2}m\dot{s}^2) \tag{A3.29}$$

work done = change of kinetic energy.

Potential. In a gravitational field $F = mg$. If z measured in reverse direction to gravitational attraction then

$$W = \int \mathbf{F} \cdot d\mathbf{s} = \int_0^z mg\,dz = mgz$$

work done = change of potential energy $\tag{A3.30}$

Power

Power is the rate of doing work $= \dot{W}$

$$= \frac{d}{dt} \int \mathbf{F} \cdot d\mathbf{s}.$$

If F is constant and angle θ is zero then

$$\dot{W} = F \frac{ds}{dt} = FV.$$

For an engine brake $V = \omega r$ and $F = T/r$

$$\dot{W} = T\omega = \frac{2\pi N T}{60}$$

or $\hspace{9cm}$ (A3.31)

$$\dot{W} = \frac{2\pi N F r}{60} = \frac{FN}{k}$$

where k is the brake constant $= 60/2\pi r$; N is the speed in rev min^{-1}; T is the torque in Nm; $\dot{W}$ is the power in W (or J s^{-1}); F is the load in N.

Mohr stress circle

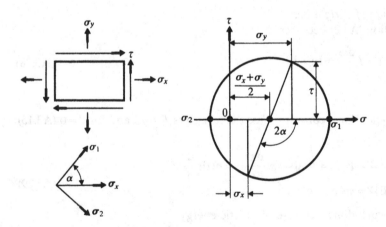

$$\sigma_{1,2} = \left(\frac{\sigma_x + \sigma_y}{2}\right) \pm \sqrt{\left[\left(\frac{\sigma_y - \sigma_x}{2}\right)^2 + \tau^2\right]},$$

$$\tan 2\alpha = -\frac{2\tau}{\sigma_y - \sigma_x}.$$

(A3.32)

Elastic bending of beams

$$\frac{M}{I} = \frac{\sigma}{y} = \frac{E}{R},$$ (A3.33)

where σ = bending stress at distance y from neutral axis; M = bending moment at section considered; R = change in radius of curvature of beam.

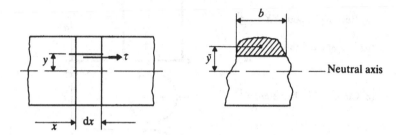

Shear stress distribution at distance x along the beam

$$\tau = \frac{VA\bar{y}}{Ib},$$ (A3.34)

where τ = shear stress at distance y from neutral axis; V = shear force at distance x along the beam; A = shaded area; $\bar{y}$ = distance of centroid of area A from the neutral axis.

Torsion formulae

Round shafts

$$\frac{T}{J} = \frac{G\theta}{l} = \frac{\tau}{r},$$ (A3.35)

where T = torque; J = polar second moment of area; θ = angle of twist; τ = shear stress at radius r.

Thin walled tube

$$\tau = \frac{T}{2At}$$ (A3.36)

and

$$\frac{\theta}{l} = \frac{T}{4A^2G} \oint \frac{ds}{t},$$ (A3.37)

where A = enclosed area; t = wall thickness; ds = short length of perimeter.

Table A3.3. *Section properties*

Second moment of area

(a) Rectangle about XX $\dfrac{bd^3}{3}$

(b) Rectangle about CC $\dfrac{bd^3}{12}$

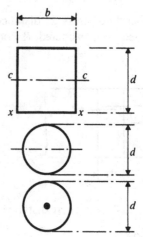

(c) Circle about diameter $\dfrac{\pi d^4}{64}$

(d) Polar moment of circle $\dfrac{\pi d^4}{32}$

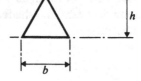

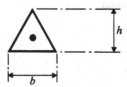

(e) Triangle about base $\dfrac{bh^3}{12}$

(f) Triangle about axis through centre of area $\dfrac{bh^3}{36}$

Parallel axis theorem $I_{XX} = I_G + Ah^2$
where XX is axis, distance h from centre G of area A
Perpendicular axis theorem $I_Z = I_X + I_Y$

Table A3.4. *Shear force and bending moment diagrams*

Beam	Shear force distribution	Bending moment distribution	Maximum deflection
W (cantilever point load)			$\dfrac{1}{3}\dfrac{Wl^3}{EI}$
w per unit length			$\dfrac{1}{8}\dfrac{wl^4}{EI}$
W (mid-span)		$\dfrac{Wl}{4}$	$\dfrac{1}{48}\dfrac{Wl^3}{EI}$
w per unit length		$\dfrac{wl^2}{8}$	$\dfrac{5}{384}\dfrac{wl^4}{EI}$
W (mid-span)		$\dfrac{Wl}{8}$, $\dfrac{Wl}{8}$	$\dfrac{1}{192}\dfrac{Wl^3}{EI}$
w per unit length		$\dfrac{wl^2}{24}$, $\dfrac{wl^2}{12}$	$\dfrac{1}{384}\dfrac{wl^4}{EI}$

Bernoulli's equation
For the steady flow of an inviscid incompressible fluid along a streamline, equation (A3.1) gives

$$\frac{p}{\rho}+\frac{V^2}{2}+gz=\text{constant.} \tag{A3.38}$$

Continuity
For steady flow, the rate of fluid mass flowing across any section of a stream tube remains constant (derived from equation (A3.3)).

$$\dot{m} = \rho AV = \text{constant.} \tag{A3.39}$$

Hydrostatic pressure

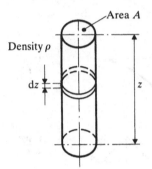

From A3.1 force F = weight of fluid column = mass × acceleration

$$= \int_0^z (\rho A \, dz)g = \rho Agz$$

Pressure due to column

$$= \frac{F}{A} = \rho gz \tag{A3.40}$$

Centre of pressure

$$y_p = \frac{A\int y^2 \, dA}{A\int y \, dA} = \frac{\text{second moment of area}}{\text{first moment of area about surface}}. \tag{A3.41}$$

Height of metacentre from centre of buoyancy

$$MB = \frac{I}{V} = \frac{\text{second moment of water line area}}{\text{total displaced volume}}. \tag{A3.42}$$

Flowmeters – orifices, nozzles and venturi meters
For incompressible flow, equations (A3.38) and (A3.39) with modification for real viscous fluids give equations of the form

$$\dot{m} = k\sqrt{h}, \tag{A3.43}$$

where k is a constant related to the dimensions of the flowmeter, viscous effects, and the densities of both the moving fluid and the manometer fluid, and h is the pressure difference (or 'head') across the flowmeter, often expressed as the length of a column of manometric fluid (see equation (A3.40)).

For compressible flow of perfect gases, equations (A3.6), (A3.39) and (A3.50) give equations of the form

$$\dot{m} = k\varepsilon\sqrt{\left(\frac{hp}{T}\right)}, \tag{A3.44}$$

where k is a constant for a particular flowmeter and manometer; h is the differential head; p is the upstream pressure; T is the upstream temperature (absolute); ε is an expansion factor used to account for the compressibility.

Pitot tubes for flow measurement
Equation (A3.38) gives $p_0 - p = \tfrac{1}{2}\rho V^2$, (A3.45)

where p_0 is the stagnation pressure; and p is the static pressure at a point where the velocity is V. Hence

$$V = \sqrt{\left[\frac{2(p_0 - p)}{\rho}\right]}$$ (A3.46)

and

$$\dot{m} = \int_{\text{area}} \rho V \, dA,$$ (A3.47)

where p_0 is the stagnation pressure in Pa; p is the static pressure in Pa; ρ is the density in kg m^{-3}; V is the velocity in m s^{-1}.

Sonic velocity
Equations (A3.1) and (A3.39) give

$$a = \sqrt{\left[\left(\frac{dp}{d\rho}\right)_s\right]}$$ (A3.48)

for liquids and gases.
   With equation (A3.6) for perfect gases (only)

$$a = \sqrt{(\gamma R T)},$$ (A3.49)

where a is sonic velocity in m s^{-1}; R is the gas constant in J kg^{-1} K^{-1}; T is the absolute temperature in K.

Steady flow energy equation
Applying equation (A3.4) to an open system with steady flow gives

$$\dot{Q} - \dot{W}_s = \dot{m} \, \Delta\left(h + \frac{V^2}{2} + gz\right).$$ (A3.50)

Entropy
Entropy is a property of a pure substance which results from equation (A3.5) such that the change in entropy

$$ds = \frac{dQ \text{ reversible}}{T}.$$ (A3.51)

Thermodynamic work
For a reversible (frictionless, quasistatic) process equation (A3.15) gives displacement work

$$W = \int p\, dV,$$ (A3.52)

where dV is the volume change.

Electric field
From equation (A3.9) add equation (A3.16),

$$\mathbf{E} = \frac{1}{4\pi\varepsilon_0} \sum_i \frac{q_i}{r_i^2} \hat{r}_i$$ (A3.53)

$$\iint_{c \cdot s} \mathbf{E} \cdot d\mathbf{S} = \sum_{c \cdot v} \frac{q_i}{\varepsilon_0} = \iiint_{c \cdot v} \frac{\rho}{\varepsilon_0} dV \quad \text{(Gauss' flux theorem)}$$ (A3.54)

From equation (A3.17)

$$\oint \mathbf{E} \cdot d\mathbf{l} = 0.$$ (A3.55)

Potential difference
From equation (A3.17)

$$V = -\int_A^B \mathbf{E} \cdot d\mathbf{l}.$$ (A3.56)

For a conducting sphere equation (A3.53) gives

$$V = \frac{Q}{4\pi\varepsilon_0 r}.$$ (A3.57)

Capacitance
From equations (A3.18) and (A3.57) for a sphere,

$$C = 4\pi\varepsilon_0 r.$$ (A3.58)

For a parallel plate condenser equation (A3.54) gives

$$E = \frac{\sigma}{\varepsilon_0} = \frac{Q}{\varepsilon_0 A}.$$ (A3.59)

Using equations (A3.56) and (A3.59) gives

$$V = \frac{Qd}{\varepsilon_0 A}.$$
(A3.60)

Thus

$$C = \frac{A\varepsilon_0}{d}.$$
(A3.61)

For capacitors in series

$$\frac{1}{C} = \sum_i \frac{1}{C_i}.$$
(A3.62)

For capacitors in parallel

$$C = \sum_i C_i.$$
(A3.63)

Resistance
From equation (A3.10)

$$R = \rho \frac{l}{A},$$
(A3.64)

where ρ = resistivity in ohm metres (Ωm).
   For resistors in series

$$R = \sum_i R_i.$$
(A3.65)

For resistors in parallel

$$\frac{1}{R} = \sum_i \frac{1}{R_i}.$$
(A3.66)

Kirchhoff's rules

(1)   From charge conservation, the algebraic sum of the currents at a junction is zero

$$\sum I = 0.$$
(A3.67)

(2)   From Ohm's law, equation (A3.10), the sum of the emfs in a circuit is equal to the sum of the voltage drops in the resistance around the circuit

$$\sum \varepsilon = \sum IR.$$
(A3.68)

Power

From equations (A3.10) and (A3.20),

$$\text{Power } \dot{W} = \frac{dW}{dt} = \frac{dW}{dq}\frac{dq}{dt} = \varepsilon I \qquad (A3.69)$$

or

$$\dot{W} = VI \quad \text{if the internal resistance of the source of emf is negligible,} \qquad (A3.70)$$

or, using equation (A3.10),

$$\dot{W} = I^2 R = \frac{V^2}{R}. \qquad (A3.71)$$

Magnetic field

From equation (A3.11)

$$\iint_{c\cdot s} \mathbf{B}\cdot d\mathbf{S} = 0 \quad \text{(Gauss' theorem for a magnetic field).} \qquad (A3.72)$$

Magnetostatics

$$\oint \mathbf{B}\cdot d\mathbf{l} = \mu_0 \sum I. \qquad (A3.73)$$

Inductance

For self-inductances in series

$$L = \sum_i L_i. \qquad (A3.74)$$

For self-inductances in parallel

$$\frac{1}{L} = \sum_i \frac{1}{L_i}. \qquad (A3.75)$$

Light

For a uniform perfectly diffusing surface

$$L = \frac{rE}{\pi} \quad \text{cd m}^{-2}\text{ (nit),} \qquad (A3.76)$$

where r is the reflection factor. Or

$$L = rE \quad \text{apostilb.} \qquad (A3.77)$$

For a lens or mirror

$$\frac{1}{u} + \frac{1}{v} = \frac{1}{f} \tag{A3.78}$$

and

$$m = \frac{v}{u}. \tag{A3.79}$$

Table A3.5. *Some standard values (four-figure accuracy)*

International Standard Atmosphere (ISA)
    Pressure = 1.013 bar = 101.3 kN m^{-2} = 101.3 kPa
    Temperature = 15 °C   = 288 K

Molal volume
    $\overline{V} = 22.41$ m^3 kg-mol^{-1} at ISA pressure and 0 °C

Composition of air

	$\overline{M}$	Vol. analysis	Grav. analysis
Nitrogen (N$_2$)	28	0.79	0.767
Oxygen (O$_2$)	32	0.21	0.233

Properties of air
    Molecular mass $\overline{M} = 29$
    Specific gas constant $R = 0.287$ kJ kg^{-1} K^{-1}
    Specific heat at constant pressure $C_p = 1.055$ kJ kg^{-1} K^{-1}
    Specific heat at constant volume $C_v = 0.718$ kJ kg^{-1} K^{-1}
    Ratio of specific heats $\gamma = C_p/C_v = 1.4$
    Thermal conductivity $k = 0.0253$ W m^{-1} K^{-1} as ISA conditions
    Dynamic viscosity $\mu = 17.9$ μPa s
    Density at ISA conditions $\rho = 1.225$ kg m^{-3}
    Sonic velocity at ISA conditions $a = 340$ m s^{-1}

Properties of water
    Molecular mass $\overline{M} = 18$
    Specific heat at constant pressure at 15 °C $C_p = 4.186$ kJ kg^{-1} K^{-1}
    Thermal conductivity $k = 595$ mW m^{-1} K^{-1} at 15 °C
    Dynamic viscosity $\mu = 1.14$ mPa s at 15 °C
    Density at 4 °C $\rho = 1000$ kg m^{-3}

Table A3.6. *Physical constants (four-figure accuracy)*

Avogadro constant	N	602.2×10^{24} kmole^{-1}
Boltzmann constant	k	13.81×10^{-24} J K^{-1}
Charge of electron	e	160.2×10^{-21} C
Magnetic permeability (free space)	μ_0	$4\pi \times 10^{-7}$ H m^{-1}
Permittivity (free space)	ε_0	8.854×10^{-12} F m^{-1}
Planck constant	h	662.6×10^{-36} Js
Standard gravity	g	9.807 m s^{-2}
Stefan–Boltzmann constant	σ	56.70 nW m^{-2} K^4
Universal gas constant	$\overline{R}$	8.314 kJ kmole^{-1} K^{-1}
Velocity of light	c	299.8×10^6 m s^{-1}

Table A3.7. *Some properties of some metals and alloys (average values at 15 °C)*

Property	Units	Aluminium	Brass	Cast iron	Copper	Lead	Mild steel	Nimonic	Silver
Density	kg m^{-3}	2710	8370	7600	8960	11 300	7800	8200	10 500
Elastic modulus	GPa	70	100	115	112	16	206	200	83
Poisson's ratio	–	0.34	0.37	0.26	0.36	0.44	0.28	–	0.37
Shear modulus	GPa	26	37	46	41	6	80	88	30
Ultimate strength	MPa	120	430	210 (T) 730 (C)	230	15	490	1200	290
Coefficient of thermal expansion	μm m^{-1} K^{-1}	24	19	9	16	29	12	12	19
Thermal conductivity	W m^{-1} K^{-1}	234	114	48	384	35	47	113	418
Melting point	K	933	1200	1500	1356	600	1600	1650	1234
Electrical resistivity	nΩm	28	62	600	17	207	160	1200	16

Notes: T = tensile; C = compressive.
These values depend on impurities, alloying elements, heat treatment, method of production, temperature, environment, etc. If required, more detailed and accurate information can be obtained from British Standard Specifications, Manufacturers Specifications, or textbooks such as *Physical and Chemical Constants* by G. W. C. Kaye & T. H. Laby (Longman).

Table A3.8. *Some properties of some non-metals (average values)*

Property	Units	Asbestos	Carbon	Concrete	Glass	Plastics Thermoplastic (polyethylene)	Thermoset (epoxy)	Wood Oak	White pine
Density	kg m^{-3}	2400	1500	2000	2500	930	1115	700	500
Elastic modulus	GPa		5		70	0.3	2.4	10.3	10.3
Poisson's ratio	–			0.10	0.25				
Coefficient of thermal expansion	μm m^{-1} K^{-1}		2.4	12	8	150	60		
Thermal conductivity	W m^{-1} K^{-1}	0.2	1.6	1.0	1.0	0.38	0.19	0.17	0.12
Electrical resistivity	TΩm				>0.01	>100	1.0		
	$\mu\Omega$m		14						

Index